כתבי האקדמיה הלאומית הישראלית למדעים

PUBLICATIONS OF THE ISRAEL ACADEMY
OF SCIENCES AND HUMANITIES

SECTION OF SCIENCES

———

FAUNA PALAESTINA

INSECTA I : DIPTERA PUPIPARA

Hippobosca equina

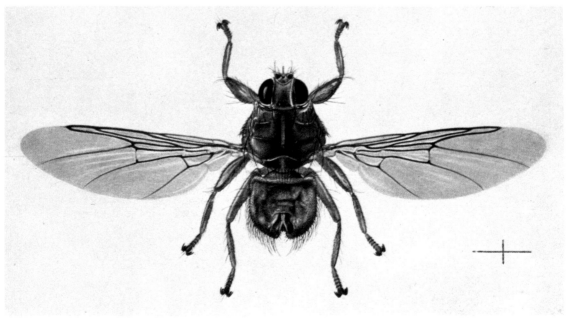

Ornithomyia avicularia

FAUNA PALAESTINA · INSECTA I

DIPTERA PUPIPARA

by

OSKAR THEODOR

Jerusalem 1975

The Israel Academy of Sciences and Humanities

Author's Address:
Institute of Microbiology
The Hebrew University, Jerusalem, Israel

Printed in Israel
at The Jerusalem Post Press, Jerusalem

CONTENTS

INTRODUCTION

THE SECTION PUPIPARA of the Diptera Cyclorrhapha consists of three families: Hippoboscidae, Streblidae and Nycteribiidae. All are blood-sucking, pupiparous, obligatory ectoparasites. The Hippoboscidae are parasites of various orders of mammals and birds, the two other families are exclusively parasites of bats.

All Pupipara have become more or less modified by adaptation to an ecto-parasitic life. Hippoboscidae and Streblidae contain normally winged forms; forms with more or less reduced, non-functional wings; forms with functional wings which break off near the base when the host is reached, and completely apterous forms which have also lost the halteres. The Nycteribiidae have lost the wings completely, and the thorax is so greatly modified that the homologies of the various parts are difficult to establish (Schlein, 1970).

The integument of all Pupipara is tough and leathery, and the segmentation of the abdomen has been lost to a great extent; it is mainly membranous and very extensible in adaptation to blood-sucking and pupipary.

The larvae develop in the uterus and are nourished by the secretion of a special gland ('milk gland'). They pass through three stages and are deposited ready to pupate. The term pupipary is thus not exact; in fact, a mature larva or praepupa is deposited. The praepupae of most Hippoboscidae and of some Streblidae are dropped to the ground. The praepupae of the wingless genus *Melophagus* are attached to the hairs of the host, and the praepupae of the Nycteribiidae and of most Streblidae, except of *Ascodipteron*, are attached to the substrate in the environment of the hosts, on the walls of caves or on leaves or branches of trees.

The Pupipara are distributed throughout the world, but most species occur in the tropics. There are eighteen genera of Hippoboscidae in the Old World, of which eleven also occur in America (two of them introduced in recent times) and three are distributed only in America. There are four genera of Streblidae in the Old World and twenty-three in America. Only one of the eleven genera of Nycteribiidae occurs both in the Old World and in America, and one genus occurs only in America.

The Pupipara show varying degrees of host–parasite specificity. A number of species are strictly species-specific or occur on several related species of a genus. Others occur on several genera of a family, and still others are quite unspecific and occur on hosts of several families or even orders. Narrow host specificity is apparently genuine in some cases, but it seems to be due mainly to ecological factors or to geographical isolation in most cases.

The Hippoboscidae are closely related to the Glossinidae and thus belong to the Cyclorrhapha calyptrata. The systematic position of the Streblidae

1

and Nycteribiidae is still under discussion. Most authors consider the Pupipara as a monophyletic group and regard the characters that distinguish the Streblidae and Nycteribiidae from the Hippoboscidae as adaptive, due to life in caves (e.g., reduction or loss of eyes) and to parasitism on bats. Others think that Streblidae and Nycteribiidae are derived from Acalyptrata. This view is mainly based on the wing venation, which resembles that of some Acalyptrata closely, and on other characters of the skeleton and musculature of the thorax.

Some Pupipara are intermediate hosts of parasitic Protozoa. *Pseudolynchia* and other genera of Hippoboscidae transmit *Haemoproteus* of birds. *Melophagus ovinus* transmits *Trypanosoma melophagium* Hoare of sheep in Europe, and *Lipoptena capreoli* transmits *Trypanosoma theodori* Hoare of goats in Israel. Species of *Penicillidia* (Nycteribiidae) transmit species of *Polychromophilus* of bats.

Some Hippoboscidae are parasitized by mites of the genera *Myialges* (Myialgesidae) and *Microlichus* (Epidermoptidae, Sarcoptiformes). Thus, *Myialges anchora* is a common parasite of *Pseudolynchia canariensis* in Israel (Plate). Females of the mite with attached groups of eggs are found mainly on the abdomen. Species of *Microlichus* are mainly found on the wings. Mallophaga are sometimes found attached to Hippoboscidae. This is not parasitism, but phoresy, as the Mallophaga use the flies only as means of transport.

Anthrax bacilli have been found in *Hippobosca rufipes* in South Africa, and *Hippobosca* may well play a role in the mechanical transmission of anthrax in horses.

A parasitic nematode of dogs, *Dipetalonema*, may be transmitted by *Hippobosca longipennis*.

The figures in brackets which appear after the locations refer to the areas on the map.

All the material dealt with in this volume is deposited in the Department of Parasitology of the Hebrew University of Jerusalem.

The following descriptions are given in shortened form, chiefly giving characters important for identification. More detailed descriptions, descriptions of Palaearctic species included in the keys which do not occur in Israel, further information on morphology, biology, etc. and full documentation will be found in the following publications:

HIPPOBOSCIDAE.—Bequaert, 1942; 1953–1957; Jobling, 1926; Maa, 1963; 1966; 1969; Theodor & Oldroyd, 1964.

STREBLIDAE AND NYCTERIBIIDAE.—Jobling, 1928; 1929; 1939; Maa 1971; Theodor, 1954; 1967 (contains a revision of the family Nycteribiidae); 1968; Theodor & Moscona, 1954.

A number of changes have been made in the terminology of various parts of the thorax after the detailed study of the thorax of the Pupipara by Schlein (1970).

All figures have been drawn from material collected in Israel, except for species which have not been recorded from Israel, but are likely to occur.

2

1

2

3

Figs. 1–2: Abdomen of *Ornithomyia chloropus* with attached Mallophaga [*Sturnidoecus sturni* (Schrank)]

Figs. 3–5: *Pseudolynchia canariensis.* 3. Head with females and groups of eggs of *Myialges anchora* Sergent & Trouessart; 4–5. Abdomen with females and groups of eggs of *Myialges anchora*

4

5

Introduction

TECHNIQUE

Hippoboscidae should be pinned, but some specimens should be kept in 70% alcohol for the study of the abdomen, which shrinks in dry material; this also facilitates dissection of the genitalia. Nycteribiidae and Streblidae are kept in 70% alcohol. Some specimens should be mounted in Canada balsam after maceration in KOH. Care should be taken to avoid dorso-ventral compression. The genitalia are mounted in Canada balsam in side view.

TERMINOLOGY

In a comparative study of the genitalia of *Calliphora, Glossina* and the Pupipara (Schlein & Theodor, 1971), the homologies of the various parts of the genitalia were established and changes in terminology were proposed. The descriptive terms used in the past are added in brackets in the introductions to the families and in the diagnoses of the genera. Some of the descriptive terms have been retained.

The more important changes are as follows:

Old Term	New Term
Hippoboscidae	
Lateral processes or plates	Praegonites
Parameres	Postgonites
Anal sclerite	Posterior part of sternite 9
Streblidae	
Nycteriboscinae:	
Digitiform processes	Praegonites
Parameres	Postgonites
Anal sclerite	Posterior part of sternite 9
Phallobase	Hypandrium
Ascodipterinae:	
Claspers	Praegonites
Parameres	Postgonites
Nycteribiidae	
Claspers	Cerci
Parameres	Praegonites
Basal arc	Hypandrium
Phallobase	Ventral process of aedeagus apodeme

ACKNOWLEDGEMENTS

We are grateful to the following institutions and publishers for permission to reproduce the following figures:

The Trustees of the British Museum (Nat. Hist.): Nycteribiidae — Figs. 214–220, 229, 288, 328, 329, 333 and the colour plate of *Hippobosca equina* and *Ornithomyia avicularia* by A. Terzi.

The Royal Entomological Society of London: Streblidae — Figs. 159, 162, 185–186, 188, 190–195, 200–206, 211–212.

Prof. E. Lindner, editor of *Die Fliegen der Palaearktischen Region*, Verlag Schweizerbarth, Stuttgart (1964): most of the figures of Hippoboscidae, except for some figures which were prepared for this volume.

The editor of *Parasitology*, Cambridge: figures from papers by Jobling, as indicated in the respective legends.

We would also like to thank R. Julius, N. Schneider and A. Sussman of the Publications Department of the Israel Academy of Sciences and Humanities for their help in preparing the manuscript for publication.

Family HIPPOBOSCIDAE Samouelle, 1819

The Entomologists' Useful Compendium, p. 302.

INTRODUCTION

MORPHOLOGY

Head: The head of the Hippoboscidae may be best interpreted as a dorso-ventrally flattened head of Muscidae, in which the mouth parts are displaced anteriorly so that the head has become prognathous. It is rounded and more or less freely movable in the Hippoboscinae, and nearly triangular ventrally and deeply fitted into the thorax in the Lipopteninae, so that lateral movement is markedly restricted. Head and thorax form a functional unit during movement among hairs and feathers of the host (Fig. 1). The dorsal surface of the head is divided into frons and vertex by the frontal suture. The frons consists of the anterior frons and the lunula, which are either separated by a suture or not. The vertex consists of an anterior membranous part, the mediovertex, and a posterior sclerotized part, the postvertex, which bears the ocelli, if present. The anterior frons forms two processes, the frontal horns, in all Hippoboscinae, except in *Ornithoica* and in the Lipopteninae. Ventral side of head with some long setae, the jugular setae, which correspond to the vibrissae of Muscidae, and a row of setae or spines at the posterior process in most Hippoboscinae (Figs. 2–4).

The antennae (Figs. 5–7) are situated in deep pits at the sides of the frons, so that they project only little above the surface in all Hippoboscidae, except in *Ornithoica,* in which they are situated close together in a common pit in the middle of the frons. The basal segment of the antennae (scape) is either free or completely or partly fused with the head capsule. The second segment (pedicel)

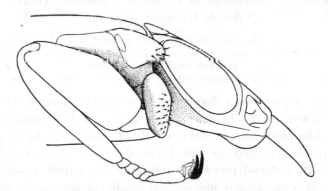

Fig. 1: *Pseudolynchia canariensis* (Macquart). Head and thorax during forward movement

5

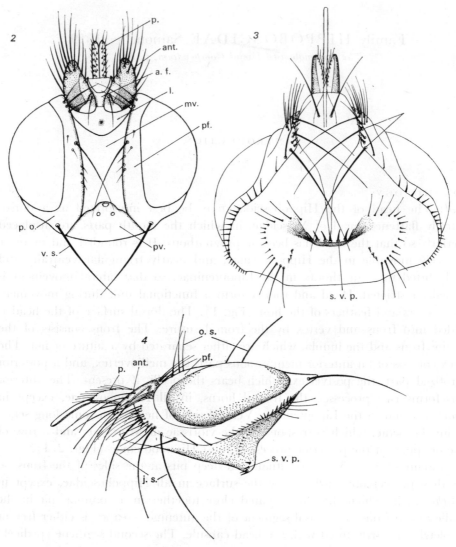

Figs. 2–4: *Ornithomyia avicularia* (Linnaeus). Head.
2. dorsal; 3. ventral; 4. lateral

ant. – antenna; a.f. – anterior frons; j.s. – jugular setae; l. – lunula;
mv. – mediovertex; o.s. – orbital setae; p. – palps; pf. – parafrontalia;
p.o. – posterior orbit; pv. – postvertex;
s.v.p. – spines at posterior ventral process; v.s. – vertical setae

is either rounded or possesses a dorsal process of varying length. This process is present in all genera of the Hippoboscinae, except in *Hippobosca* and in the Lipopteninae. The antennal pits have a distinct dorsal margin in *Hippobosca* and in the Lipopteninae and are called 'closed'. There is no dorsal margin in the antennae with a dorsal process and the pits are called 'open'. The third antennal segment is completely invaginated inside the second, so that only the arista projects from a small opening. The antennae thus appear 1-segmented. The arista is branched in *Hippobosca,* in *Lipoptena* and in *Melophagus,* and is

6

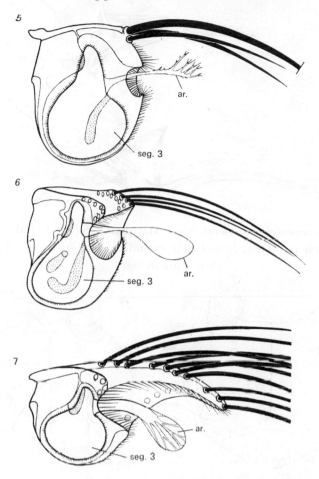

Figs. 5–7: Antenna. 5. *Hippobosca equina* (Linnaeus). 6. *Pseudolynchia canariensis* (Macquart). 7. *Ornithomyia avicularia* (Linnaeus)

ar. – arista; seg. 3 – segment 3

spatula-shaped or leaf-shaped in the Ornithomyiini (Figs. 5–7). The maxillary palps are 1-segmented and vary in length. They project anteriorly and enclose the mouth parts in some genera.

The structure of the proboscis resembles that of blood-sucking Muscidae, for example, *Glossina*, but the proboscis is retracted into the head capsule when not in use (Figs. 8–9). Both sexes have similar mouth parts, as they both suck blood. The structure of the head and mouth parts has been studied in detail by Jobling (1926).

Thorax: The thorax (Figs. 10-12) is more or less dorso-ventrally flattened. The sternal plate is very wide and the legs are inserted at the sides. Some pleural elements are displaced dorsally. For example, the notopleuron and the dorsal part of the mesopleuron are visible dorsally and the anterior spiracle lies on the

7

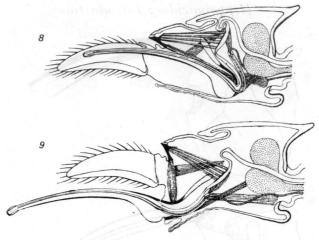

Figs. 8–9: *Pseudolynchia canariensis* (Macquart). Sagittal section of head.
8. proboscis retracted; 9. proboscis extended (modified after Jobling,
1926, *Parasitology*)

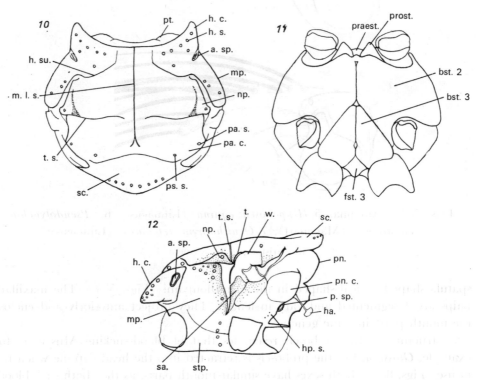

Figs. 10–12: *Ornithomyia avicularia* (Linnaeus). Thorax.
10. dorsal; 11. ventral; 12. lateral

a.sp. – anterior spiracle; bst. 2, 3 – basiternum 2, 3; fst. 3 – furcasternum 3; ha. – haltere;
h.c. – humeral callus; h.s. – humeral setae; h.su. – humeral suture; hp.s. – hypopleural setae;
mp. – mesopleuron; m.l.s. – median longitudinal suture; np. – notopleuron;
pa.c. – postalar callus; pa.s. – postalar setae; pn. – postnotum; pn.c. – postnotal callus;
praest. – praesternum; prost. – prosternum; p.sp. – posterior spiracle;
ps.s. – praescutellar seta; pt. – prothorax; sa. – subalifer; sc. – scutellum;
stp. – sternopleuron; t. – tegula; t.s. – transverse suture; w. – wing

dorsal surface in some species. The humeral calli are either low and rounded (*Hippobosca, Ornithoica,* Lipopteninae) or are conical and project markedly anteriorly, and the head rests on them during forward movement. The median longitudinal suture is well developed in most genera, but is absent in some, for example, in *Hippobosca* and *Ornithoica.* The humeral and the scutellar sutures are always distinct, the notopleural suture may be distinct or obliterated. The transverse suture of the mesonotum may be complete or obliterated in the middle. In the Lipopteninae it consists of two halves which are directed posteriorly and inwards, and it is absent in *Melophagus.* The sternal plate consists of three segments: The prosternum, which is divided into two triangular processes in most genera, but undivided in *Hippobosca* and *Ornithoica.* A small triangular plate with setae, the praesternum, is situated in front of the prosternum. Basisterna 2 and 3 are divided by a median longitudinal suture. Furcasterna 1 and 2 have been lost and only furcasternum 3 is present. The pleurae resemble those of the higher Muscidae in general (Fig. 12). The anterior spiracle is rounded or oval and situated at the base of the humeral callus, directed laterally in most species, but on the dorsal surface in some species. The lateral protuberances of the postnotum, the postnotal calli, vary markedly in form in the various genera. They may be low or markedly projecting and bear processes of varying form in some genera. They may bear one or two rows of setae; in some genera they are bare.

Wings: The wings are normally developed in most species. The membrane is tough in genera with permanently functional wings, delicate in genera which lose the wings. The wings are reduced in size and non-functional in three genera. Some genera of Lipopteninae have functional wings when they hatch, but these break off near the base when the host is reached. *Melophagus* has lost wings completely, except for small, sclerotized stumps which are apparently rudiments of the basal, thickened part of the costa (Figs. 13–14). Halteres are normally developed in all species, except in *Melophagus,* in which they are absent. The wings

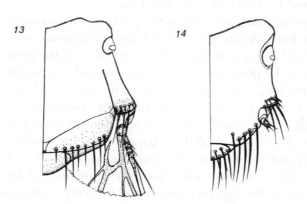

Figs. 13–14: Wing rudiments. 13. *Lipoptena;* 14. *Melophagus*

9

are folded flat above the abdomen at rest. The venation of the wings in genera with permanently functional wings resembles that of the higher Muscidae (Figs. 15–16). It differs mainly in the absence of a discal vein, so that there is no

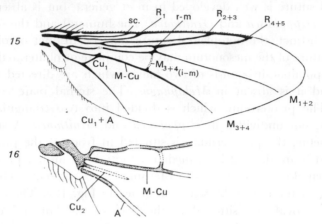

Figs. 15–16: Wing. 15. *Ornithomyia avicularia* (Linnaeus); 16. *Icosta*, basis of wing with rudiment of Cu_2

discal cell, and in the always straight $M_1 + _2$. The costa is sclerotized to the end of $R_4 + _5$. The branches of the radius are strongly sclerotized and usually pigmented; the posterior veins are weakly developed distal to the cross veins, and the anterior veins are concentrated near the costa, so that the posterior part of the wings seems to be without distinct venation. There may be one, two or three cross veins. The middle cross vein (i–m, $M_3 + _4$) is situated near the anterior cross vein or distinctly proximal to it. The membrane is covered with microtrichia to a varying extent in most genera, but microtrichia are absent in *Hippobosca* and *Ornithophila*. The wings of the Lipopteninae have only three well marked longitudinal veins, R_1, $R_4 + _5$ and $M_3 + _4$ (Fig. 141). The other veins are reduced to folds in the membrane. There is only one cross vein, which is apparently formed by fusion of i–m and r–m.

Legs: The legs are short and thick in the parasites of mammals, longer and more slender in the bird parasites. Coxa 1 is large, cushion-shaped and serves as support for the head together with the humeral calli and the prothorax during forward movement. Its posterior surface is concave and rests on the femur (Fig. 1). Praetarsus with long claws, pulvilli and a feathered or bare empodium. One or both pulvilli may be rudimentary. The claws consist of a basal part and a distal part which is recurved at an acute angle on the basal part. Such claws are called simple. The claws of all bird parasites, except *Ornithoica*, are divided, so that they appear trifid in profile if the basal part is included. They are called double claws (Figs. 17-18). The claws are asymmetrical in some species of *Hippobosca* and in the Lipopteninae, one claw being shorter and more sharply curved (Fig. 155). The pulvilli are also of different size in these cases; the larger pulvillus stands near the shorter claw.

10

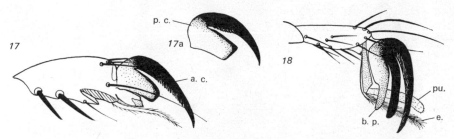

Figs. 17–18: Praetarsus.   17. *Melophagus ovinus* (Linnaeus), simple claw;
17a. *M. ovinus,* posterior claw; 18. *Stenepteryx,* double claw
a.c. – anterior claw; b.p. – basal part; e. – empodium; p.c. – posterior claw (if the leg
is extended laterally and horizontally); pu. – pulvillus

Abdomen: The abdomen is membranous in its greater part (Figs. 19–24). Its integument consists mainly of endocuticula and the exocuticula is very thin, except on the tergal plates, which are usually reduced in size. The abdomen may increase markedly in size by gravidity, growth of the intestine and blood-sucking, particularly in the Lipopteninac and in the Hippoboscinae with reduced wings. It may become several times the size of the thorax, while it is smaller than the thorax in freshly hatched specimens. Only the length of head + thorax

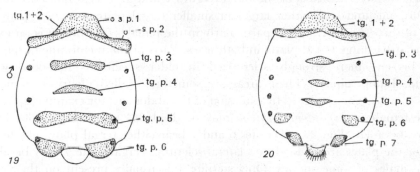

Figs. 19–20: *Hippobosca equina* Linnaeus. Abdomen, dorsal.
19. male; 20. female

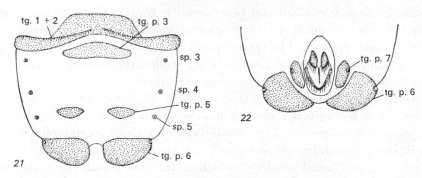

Figs. 21–22; *H. camelina* Leach. Male abdomen.
21. dorsal; 22. ventral, posterior part

11

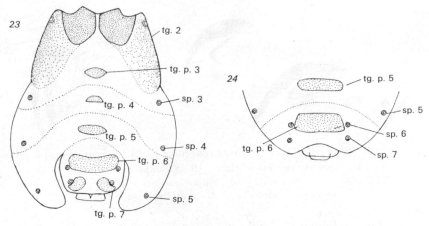

Figs. 23–24: *Lipoptena capreoli* Rondani. Abdomen, dorsal.
23. female; 24. male, posterior part
sp. – spiracle; tg. – tergite; tg.p. – tergal plate

is, therefore, given in the descriptions. There are seven pairs of abdominal spiracles, but the limits of the segments are more or less obliterated by membranization and displacement. Only the basal tergite is always well developed and sclerotized. It consists of the fused tergites 1 and 2. Tergites 3–5 are usually membranous and bear either larger or smaller tergal plates, or these plates, or some of them, may be absent, particularly in the female. *Ornithoica* is an exception, as it has large tergal plates in both sexes. Parts of the membrane of tergite 3 may become more strongly sclerotized in older specimens in some genera (*Icosta*, Lipopteninae). These areas are sometimes called pleurites. Some spiracles may become displaced by extension of the abdomen; for example, spiracle 5 of the female of *Lipoptena capreoli* may be displaced into the posterior bulges of the abdomen (Fig. 23). Tergites 6 and 7 bear either tergal plates like tergites 3–5, or the plates are divided into lateral sclerites. Tergal plate 7 may be absent in the males of some species. Only sternite 1 is usually present on the ventral surface in most genera, but it is absent in *Icosta* and *Pseudolynchia*. Sternite 5 bears large paired plates in the males of some genera (*Allobosca, Ornithophila*). The anal frame is sclerotized and bears setae; it is open ventrally in the male. It may be closed in the female and bears plates at the ventral margin which cover the genital opening. The ventral part of the anal frame continues into the genital opening. This ventral part, which is called the genital sclerite, varies markedly in form and may bear processes and spines.

Male Genitalia: The genitalia (Fig. 25) are of similar structure in all species examined. The aedeagus is a more or less conical tube which is sclerotized dorsally and sometimes laterally. The ventral membrane forms an endophallus which varies in size in the different species and may bear spines. The aedeagus articulates with a plate-shaped apodeme, which is articulated with the hypandrium. Hypandrium and postgonites (parameres) are fused into a rigid sclerite. The postgonites bear a number of small, tubular sense organs (Fig. 26). The genitalia

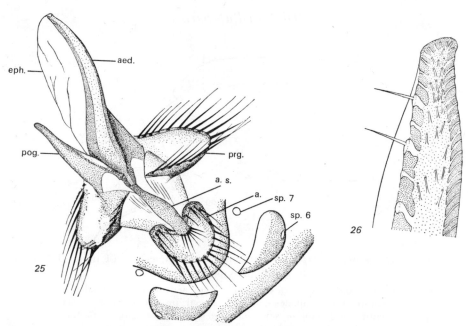

Fig. 25–26: *Hippobosca longipennis* Fabricius. 25. male genitalia, extended;
26. end of postgonite with sense organs

a. – anus; aed. – aedeagus; a.s. – anal sclerite; eph. – endophallus; pog. – postgonite;
prg. – praegonite; sp. – spiracle

are retracted into the abdomen at rest and are protruded by muscles during copulation (Figs. 27–28). The external genitalia are reduced to two small processes
near the genital opening, which are folded laterally during copulation. Their form
varies in the different genera. They were called lateral processes or plates in the
past, but are apparently the praegonites (Figs. 29–36). A small, oblong sclerite,
the anal sclerite (posterior part of sternite 9), connects the posterior dorsal end
of the postgonites (parameres) with the anal frame and restricts the extension of
the genitalia. Its form varies in the different genera and it is apparently absent
in some genera.

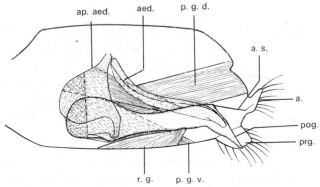

Fig. 27: *Hippobosca longipennis* Fabricius. Sagittal section of male
abdomen, genitalia retracted
(See Fig. 28 for explanation of abbreviations)

13

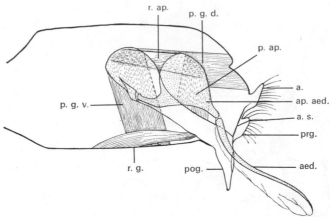

Fig. 28: *Hippobosca longipennis* Fabricius. Sagittal section of male abdomen, genitalia extended

a. – anus; aed. – aedeagus; ap.aed. – apodeme of aedeagus; a.s. – anal sclerite; p.ap. – protractor of apodeme; p.g.d. – dorsal protractor of genitalia; p.g.v. – ventral protractor of genitalia; pog. – postgonite; prg. – praegonite; r.ap. – retractor of apodeme; r.g. – retractor of genitalia

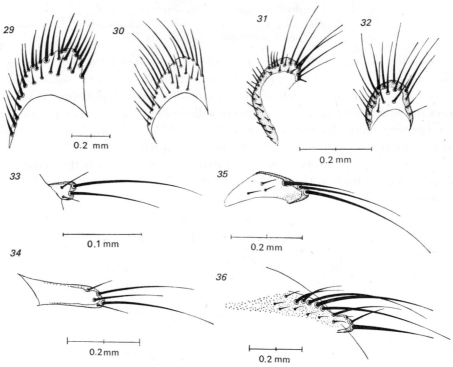

Figs. 29–36: Praegonites.
29. *Hippobosca camelina* Leach; 30. *H. longipennis* Fabricius; 31. Pseudolynchia *canariensis* (Macquart); 32. *Lipoptena capreoli* Rondani; 33. *Ornithoica vicina* (Walker); 34. *Ornithophila metallica* (Schiner); 35. *Ornithomyia avicularia* (Linnaeus); 36. *Crataerina pallida* Latreille

14

Chaetotaxy: The chaetotaxy of the Hippoboscidae is less constant than in other higher Muscidae. The number, length and position of the setae vary markedly in the same species, and the setae may be reduced or absent on areas exposed to strong friction, for example, on the sternal plate of the thorax. There are often marked differences in the chaetotaxy of males and females.

SEXUAL DIMORPHISM

The differences between males and females consist mainly in the form, size and number of the tergal plates of the abdomen or in their absence in one sex. Thus, tergal plate 7 is absent in the males of some species, or tergal plates 3–5 may be absent in the females. Chaetotactic differences between the sexes are common and often marked (*Hippobosca, Ornithoica*). The form of the scutellum is different in both sexes in some species (*Hippobosca camelina*).
Gynandromorphs have been found to be common in some species of *Ornithoica*.

COLORATION

Most Hippoboscidae are brown or black. There are distinct colour patterns in a few genera, for instance, the white spots on the mesonotum of *Hippobosca* or the dark spots on the ventral side of the head and thorax in species of *Ornithomyia*. A number of species, for example, in the genera *Ornithomyia* and *Crataerina,* show a distinct greenish coloration (frontispiece). This is caused not by a pigment but by the colour of the haemolymph. Kemper (1951) states that this colour is produced by the decomposition of haemoglobin, which forms a so-called verdoglobin.

DEVELOPMENT

The larva develops in the uterus through three stages. It is white when deposited, except for the black spiracular plate; it becomes dark in a few hours. It is not capable of movement like the larva of *Glossina*. The puparium is ovoid, slightly flattened dorso-ventrally. The spiracles of the puparium consist of a 3-branched atrium with numerous secondary branches (Figs. 37–38). These branches are apparently homologous to the three spiracular slits of Muscidae. Only *Melophagus* has rounded spiracles, which are situated inside or near a pit (Fig. 39). The pupal stage lasts relatively long, 20–30 days at 30° C.
The reproduction potential of Hippoboscidae is low, and a female may produce only ten to twelve larvae. The flies may live a long time, four to six months, but freshly hatched flies die in a few days if they have no opportunity to suck blood.
The midgut of Hippoboscidae is long and tubular and apparently not adapted to store a large quantity of blood as it is in bed bugs and mosquitoes. Hippoboscidae feed frequently, but ingest only small quantities of blood, like other ectoparasites which spend most of their life on the host, lice for example. The midgut increases considerably in length during the life of the individual. The midgut of a freshly hatched *Lipoptena capreoli* is 1 cm long, that of a fly fourteen to twenty days old 4–5 cm. This growth is effected by increase in size of the cells; mitoses have not been found (Theodor, 1928).

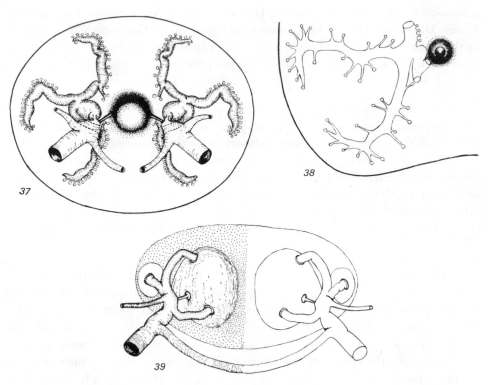

Figs. 37–39: Spiracles of puparium. 37. *Lipoptena capreoli* Rondani;
38. *Hippobosca equina* Linnaeus; *39. Melophagus ovinus* (Linnaeus)

The apodemes of the endoskeleton of the thorax and of the male genitalia increase markedly in size during the life of the individual. This increase is effected by deposition of skeletal material by active epithelium at the end of the apodemes (Figs. 40–41). The muscles of the thorax of *Lipoptena capreoli* grow markedly

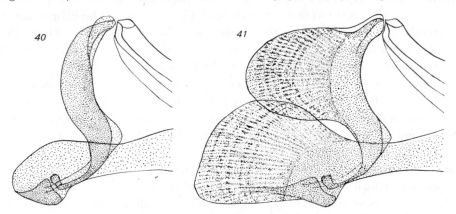

Figs. 40–41: *Hippobosca longipennis* Fabricius, male. Apodemes of
aedeagus and sternite 9. 40. freshly hatched specimen;
41. older specimen

by increase in size of the fibres and by production of new fibres. They also change their position and occupy the place vacated by the indirect flight muscles which are histolysed after the wings are lost (Schlein, 1967).

DISTRIBUTION

About one hundred and ten species of the about one hundred and fifty species known are distributed in the Old World; thirty-six are found only in America. Eight genera, including all mammal parasites, except the Lipopteninae, are restricted to the Old World and three genera are restricted to America. A few species are nearly cosmopolitan, but most species have a more or less restricted distribution. Eighteen of the thirty-six species occurring in the Palaearctic region are restricted to it. The others are either cosmopolitan or tropical species which reach the region in its southern parts. One species, *Ornithomyia chloropus*, is distributed to 70° lat. N.

The fauna of Hippoboscidae in Israel and adjacent areas consists mainly of Palaearctic (Mediterranean) species. Two or three Ethiopian species occasionally enter the Mediterranean, only one of which, *Ornithoica turdi*, has been found in Israel so far.

HOST–PARASITE RELATIONSHIP

All Hippoboscidae are either strictly mammalian or strictly avian parasites; no species of Hippoboscidae occurs regularly on both mammals and birds. This is also true for the genera, with the sole exception of *Hippobosca struthionis* Janson, which belongs to a genus otherwise restricted to mammals. The species lives on the African ostrich, which is often found together with herds of antelopes and zebras. Hippoboscidae are found on only five orders of mammals; twenty-seven species on Artiodactyla, of which three also occur on Equidae, one species on carnivores, two on lemurs and several on kangaroos. All species of Lipopteninae live exclusively on Artiodactyla.

About ninety species of Hippoboscidae live on eighteen orders of birds; seventeen each on Passeriformes and Falconiformes, eleven on Strigiformes, nine each on Galliformes and Apodiformes, eight on Columbiformes and one to six on other orders.

The Hippoboscidae show a wide variation of host specificity from strict species specificity to occurrence on species of different orders. Most species-specific species are either apterous, have reduced wings or lose their wings.

Host specificity is more marked in the mammal parasites than in the bird parasites. The degree of specificity seems to be connected with specialization; the more primitive forms occur on a large number of unrelated hosts.

Diptera Pupipara

HIPPOBOSCIDAE OF DOMESTIC ANIMALS

Pseudolynchia canariensis has become distributed throughout the tropics and subtropics with the domestic pigeon. Its original host is probably *Columba livia*. It also occurs on many other species of birds in the Old World, but only on the domestic pigeon in America. *Hippobosca longipennis* has become widely distributed with the domestic dog, but not as widely as *Pseudolynchia canariensis*. Its original hosts are wild carnivores in E. Africa, including the lion, but it has not been found so far on the domestic cat. *H. longipennis* occurs in the Mediterranean, Africa and S. Asia in more-or-less arid habitats, but not in cold or humid tropical climates. *Hippobosca equina* and *H. variegata* are exclusively parasites of domestic Equidae and Bovidae. They do not occur on wild Equidae and Bovidae, and their original hosts are not known. *H. equina* is mainly Palaearctic, but also occurs sporadically in E. Africa and is widely distributed in the Oriental region. *H. variegata* is distributed in Africa south of the Sahara, mainly in arid areas, and also in the Oriental region. It reaches the Palaearctic region in N. Africa and in Iran.

Melophagus ovinus is a specific parasite of the domestic sheep and does not occur on wild sheep. It is mainly restricted to temperate latitudes and does not occur in the tropics, except in a few high localities. *Lipoptena capreoli* is a parasite of domestic goats from the E. Mediterranean to N.W. India. It has been recorded rarely from wild goats, which probably obtain it by contact with feral domestic goats, but it may have been confused with the closely related *L. chalcomelaena*, which occurs on ibex in E. Africa and the Near East.

SYSTEMATIC PART

DIAGNOSIS OF THE FAMILY HIPPOBOSCIDAE

Dorso-ventrally flattened flies with leathery integument. Size 2–10 mm.

Head: Prognathous, dorso-ventrally flattened, divided by the frontal suture into frons and vertex. Eyes large in normally winged forms, more or less reduced in size in species with reduced wings, in species which lose their wings and in wingless species. Three ocelli are well developed, reduced or absent. Antennae short, deeply sunk in pits on the frons, 3-segmented, scape free, or completely or partly fused with the head capsule, third segment completely invaginated in second (first visible) segment. Arista either branched, dendriform or spatula-shaped. Palps 1-segmented.

Thorax: More or less flattened, sternal plate very wide. Legs inserted laterally. Prothorax small, sometimes visible dorsally. Humeral calli either low and rounded or conical and strongly projecting. Median longitudinal suture present or absent. Transverse suture of mesonotum either complete or interrupted in the middle. Scutellum well developed, usually with setae. Postnotal calli either strongly

18

projecting, with one or two rows of setae, sometimes with processes, or rounded and bare.

Wings: Normally developed, with six to seven longitudinal veins and humeral cross vein and with one, two or three cross veins; discal cell open in the subfamily Hippoboscinae. Wings with only three well marked longitudinal veins in the subfamily Lipopteninae, in which the wings are normally developed at hatching and break off near the base when the host is reached. Membrane of wings either partly or completely covered with microtrichia or microtrichia absent. Wings reduced, non-functional, with reduced venation in some genera of Hippoboscinae, reduced to small, sclerotized stumps in *Melophagus*. Halteres normally developed, absent in *Melophagus*.

Legs: Femora usually swollen. Claws strongly developed, asymmetrical in some species, simple in all species living on mammals, double in all species living on birds, except in *Ornithoica*, which has simple claws. Pulvilli normally developed, reduced or absent. Empodium feathered or nearly bare.

Abdomen: Membranous, with markedly reduced segmentation. Seven pairs of abdominal spiracles. Tergal plates more or less reduced, particularly in the female. Seven tergal plates present only in a few species. Tergites 1 and 2 always fused. Ventral side of abdomen membranous, except for sternite 1 which is also absent in some genera. Sclerites on sternite 5 present in the males of a few species.

Male Genitalia: Retracted inside the abdomen at rest. Aedeagus usually with an endophallus. Apodeme of aedeagus broad, plate-shaped, articulating with the hypandrium. Hypandrium and postgonites fused into a rigid sclerite. External genitalia reduced to two praegonites (lateral processes). Cerci absent.

Twenty genera, one hundred and thirty species, distributed throughout the world.

CLASSIFICATION

Bequaert (1942; 1953) divided the Hippoboscidae into six subfamilies, following Speiser (1908). Theodor and Oldroyd (1964) recognize only two subfamilies, which are defined as follows:

HIPPOBOSCINAE

Head rounded, freely movable. Frons indented anteriorly, forming horns, except in *Ornithoica*. Eyes large or more or less reduced in species with reduced wings. Wings either normally developed, with six to seven longitudinal veins, or reduced and non-functional. Wing membrane lost in its greater part in *Allobosca* Speiser. Alula present, sometimes reduced. Transverse suture or mesonotum complete or narrowly interrupted in the middle. Claws simple or double. Parasites of mammals and birds. Seventeen genera, of which nine occur in the Palaearctic region and eight have been recorded from Israel or are likely to occur.

19

Diptera Pupipara

LIPOPTENINAE

Head triangular ventrally, trapezoidal dorsally, deeply fitted into the thorax, so that lateral movement is markedly restricted. Anterior margin of frons smoothly rounded. Transverse suture of mesonotum always broadly interrupted in the middle, the two lateral halves directed posteriorly and inwards to near the scutellar suture or suture absent (*Melophagus*). Wings and halteres absent in *Melophagus,* functional at hatching, with only three longitudinal veins, breaking off near the base when the host is reached in the other two genera. Alula absent. Claws simple. Only parasites of Artiodactyla. Three genera, of which two are Palaearctic, both represented in Israel.

The subfamily Hippoboscinae is divided into two tribes as follows:

HIPPOBOSCINI: Humeral calli low and rounded. Postnotal calli strongly projecting, with one or two rows of setae. Antennae without dorsal process. Antennal pits closed. Arista branched. Claws simple. On mammals and ostrich. Only genus: *Hippobosca* Linnaeus.

ORNITHOMYIINI: Humeral calli conical, rounded only in *Ornithoica*. Postnotal calli rounded, more or less projecting or with processes, with a row of setae or bare. Antennae with dorsal process, antennal pits open. Arista spatula-shaped. Claws simple in the mammal parasites *Allobosca* Speiser, *Proparabosca* Theodor and Oldroyd, *Ortholfersia* Bequaert and *Austrolfersia* Bequaert and in *Ornithoica*, which lives on birds, double in all other genera. Only on birds in the Palaearctic region; *Allobosca* and *Proparabosca* on lemurs in Malagasy; *Ortholfersia* and *Austrolfersia* on kangaroos in Australia.
Sixteen genera, of which eight occur in the Palaearctic Region and three occur only in America.

Key to the Palaearctic Genera of Hippoboscidae

1. Head smoothly rounded anteriorly, triangular ventrally, deeply fitted into the thorax, so that lateral movement is restricted. Eyes reduced, situated at the anterior margin of the head, widely separated from the posterior margin. Wings and halteres absent or normally developed at hatching, breaking off near the base when the host is reached. Only three longitudinal veins. Parasites of Artiodactyla (subfamily Lipopteninae)   2
   – Head rounded, with deeply indented anterior margin (except in *Ornithoica*), freely movable. Eyes either large or reduced and widely separated from both the anterior and posterior margin of the head. Wings either normally developed, with six or seven longitudinal veins and alula or reduced and nonfunctional. Parasites of mammals and birds (subfamily Hippoboscinae)   3

20

2. Wings reduced to small, sclerotized stumps with a few setae. Halteres absent. Mesonotal transverse suture absent. **Melophagus** Latreille
 – Wings normally developed at hatching, with three longitudinal veins. Halteres normally developed. Mesonotal suture present, divided into two halves, which are directed posteriorly and inwards. **Lipoptena** Nitzsch

3. Humeral calli broadly rounded. Postnotal calli strongly projecting. Antennae without dorsal process. Arista branched. Antennal pits closed. Claws simple. Wing membrane wrinkled, without microtrichia. On mammals and ostrich (tribe Hippoboscini). **Hippobosca** Linnaeus
 – Humeral calli conical, strongly projecting (except in *Ornithoica*). Postnotal calli either rounded or more or less projecting or with processes. Antennae with dorsal process. Antennal pits open. Arista spatula-shaped. Claws simple in *Ornithoica*, double in all other genera parasitic on birds. Only on birds in the Palaearctic region (tribe Ornithomyiini) 4

4. Wings reduced, non-functional 5
 – Wings normally developed, permanently functional 6

5. Head rounded, eyes half as long as the head. Ocelli present, sometimes rudimentary. Wings very narrow, about eight times as long as wide. **Stenepteryx** Leach
 – Head oblong-oval. Eyes one-third as long as the head. Ocelli absent. Wings wider, two to four times as long as wide. **Crataerina** Olfers

6. Wings with closed anal cell, that is, with three cross veins (humeral vein not included). Ocelli always well developed 7
 – Anal cell open, that is, with only one or two cross veins. Ocelli absent in most species, distinct or rudimentary in a few species 9

7. Antennae contiguous, situated in a common pit in the middle of the frons. Humeral calli low, rounded. Claws simple. Prosternum undivided. Tergal plates of abdomen very large in both sexes. **Ornithoica** Rondani
 – Antennae widely separated, situated in pits at the sides of the frons. Humeral calli conical, strongly projecting. Claws double 8

8. Antennae with large dorsal process. Costa interrupted between Sc and R_1. R_{2+3} nearly contiguous with the costa in its apical part. Anterior spiracle situated on dorsal surface of mesonotum at base of humeral callus. Postnotal calli with a vertical row of setae. **Ornithophila** Rondani
 – Antennae with a shorter dorsal process. Costa interrupted before ending of subcosta, R_{2+3} forming a distinct angle with the costa. Anterior spiracles directed laterally, situated at the sides of the thorax. Postnotal calli bare. **Ornithomyia** Latreille

9. Wings with only one cross vein (r–m). Scutellum nearly rectangular, with finger-shaped processes in the lateral parts of the posterior margin. **Pseudolynchia** Bequaert
 – Wings with two cross veins 10

10. Head with three rounded bulges at the posterior margin. Frons nearly as long as vertex, including the postvertex, which reaches nearly to the frontal suture. Lunula fused with anterior frons without suture. Frontal horns broad, striated, with laterally directed point, only separated by a narrow groove. Postnotal calli with a posterior process, but without long setae. Scutellum without setae. Mainly parasites of oceanic birds. **Olfersia** Wiedemann

 — Postvertex small, triangular and separated from the frons by the long, membranous mediovertex. Frontal horns slender, with a rounded indentation between them. Lunula separated from the frons by a suture. Postnotal calli low, rounded, with a row of setae. Scutellum with two long setae. **Icosta** Speiser

Some species that occur in the E. Mediterranean and have not been recorded from Israel, but may be introduced by migrating birds, have been included.

Subfamily HIPPOBOSCINAE Speiser, 1908

DIAGNOSIS

Head rounded, freely movable. Frons indented anteriorly, forming horns, except in *Ornithoica*. Eyes large or more or less reduced in species with reduced wings. Wings either normally developed, with six to seven logitudinal veins, or reduced and not functional. Wing membrane lost in its greater part in *Allobosca* Speiser, Alula present, sometimes reduced. Transverse suture of mesonotum complete or narrowly interrupted in the middle. Claws simple or double. Parasites of mammals and birds. Seventeen genera, of which nine are Palaearctic and eight have been recorded from Israel or are likely to occur.

Tribe HIPPOBOSCINI Theodor & Oldroyd (1964) in Lindner, 65, Hippoboscidae, p. 18.

Genus HIPPOBOSCA Linnaeus, 1758
Systema Naturae (10th ed.), p. 607

Nirmomyia Nitzsch, 1818. *Germar's Mag. Ent.,* 3 : 309.
Zoomyia Bigot, 1885. *Annls Soc. Ent. Fr.,* Ser. 6, Vol. 5, pp. 227, 234.
Struthiobosca Maa, 1963, *Pacif. Insects,* Monograph VI, pp. 80, 127.

Type Species: *Hippobosca equina* Linnaeus, 1758.
Size 5–10 mm. Coloration reddish brown to blackish brown, with a pattern of well marked white spots on the mesonotum in most species.

Head: Rounded, freely movable. Eyes large, reacting from the anterior to the posterior margin of the head, so that there are no posterior orbits. Parafrontalia strip-like, parallel-sided or wider in the middle. Postvertex triangular or rounded anteriorly. Ocelli absent. Lunula fused with frons without suture. Anterior frons with diverging horns, which bear setae anteriorly. Antennae short, rounded, without dorsal process. Antennal pits closed. Arista branched. Palps 1-segmented, short.

Thorax: Rounded. Humeral calli rounded, low, not projecting. Anterior spiracle large, elliptical, directed laterally. Humeral and transverse suture of mesonotum well developed, median longitudinal suture absent. Scutellum rectangular or rhomboidal. Postnotal calli strongly projecting, with one or two vertical rows of setae.

Wings: With strongly sclerotized, usually darkly pigmented veins in the anterior part. Only two cross veins, anal cross vein absent. Wing membrane wrinkled, without microtrichia. *Legs:* Femora markedly swollen. Claws simple, usually asymmetrical. Pulvilli also asymmetrical or rudimentary. Empodium feathered or practically bare.

Abdomen: Membranous in its greater part. Only tergite $1 + 2$ constantly well developed. Tergal plates 3–7 present in the females of some species, divided into lateral sclerites on tergites 6 and 7; tergal plate 7 absent in the males of these species; tergal plates 3–5, or some of them, absent in the females of some species; tergal plate 7 present in the males of these species. Basal sternite trapezoidal, with fine setae. Genital sclerite of female markedly differentiated in the different species.

Male Genitalia: Aedeagus with large endophallus, which is covered with spines in most species. Praegonites (lateral processes) large, rounded, with numerous setae.

Eight species, mainly in the Ethiopian region. Three species are widely distributed in the Palaearctic region and one Ethiopian species (*H. variegata* Megerle) reaches the southern parts of the Palaearctic region. Three species occur in Israel.

Key to the Palaearctic Species of Hippobosca

1. Palps shorter than frons. Prothorax without setae. R_{2+3} about as long as the part of R_{4+5} distal to the cross vein r–m. R_{2+3} ends markedly distal to R_1 in the costa. Section of costa between R_{2+3} and R_{4+5} two to three times longer than the section between R_1 and R_{2+3}. Tergal plates 3–5 of abdomen present, larger in males. Lateral sclerites of tergite 6 small in both sexes. Lateral sclerites of tergite 7 present in the female, absent in the male   2

 – Palps as long as the frons. Prothorax with a row of setae. R_{2+3} much shorter than the part of R_{4+5} distal to r–m. R_1 and R_{2+3} end at almost the same point in the costa. Tergal plates 3–5 absent, or tergal plate 4 absent and tergal plate 5 reduced to small lateral sclerites in the male, absent in the female. Lateral sclerites of tergite 6 very large and sclerites of tergite 7 present in the male   3

2. Frontal horns broad, with convex inner margin. Scutellum white in the middle, dark laterally, with five to eleven, usually eight praemarginal setae, which usually form two groups. Veins darkly pigmented, wing length 6.5–8.0 mm. Colour chestnut-brown, with well defined white spots on the mesonotum. On horses and cattle. **Hippobosca equina** Linnaeus

 – Frontal horns slender, triangular, with straight inner margin. Scutellum almost completely white, sometimes narrowly dark laterally and with 4–7, usually 5–6 setae. Veins pale, dark only near the cross veins. Smaller and paler species. Wing length 5–6 mm. On carnivores. **Hippobosca longipennis** Fabricius

3. Large species. Wing length 9–10 mm. Postvertex as long as wide at the base. Two to three vertical setae at each side. Scutellum rhomboidal, with distinct posterior angle in the female, with a light spot in the middle and with four to eight setae near the posterior margin. Scutellum of male rounded posteriorly, with a double row of ten to twenty light setae. Postnotal calli with a vertical row of setae. Claws symmetrical, pulvilli rudimentary. R_{2+3} two to three times as long as r–m. Tergal plate 3 always present. Tergal plate 5 reduced to small lateral sclerites in the male, absent in the female. Setae in posterior part of ventral side of abdomen of female not situated on projecting tubercles. Endophallus of aedeagus with spines, aedeagus with terminal hook.

 Hippobosca camelina Leach

 – Smaller species. Wing length 7–8 mm. Postvertex distinctly longer than wide at the base. Only one pair of vertical setae. Scutellum nearly rectangular, with two dark and three light spots. Postnotal calli with two rows of setae on separate ridges. R_{2+3} only little longer than r–m. Claws asymmetrical, one pulvillus normal, the other reduced on fore- and mid-legs. Tergal plates 3–5 absent in both sexes. Posterior part of ventral side of abdomen of female with short spines on large, sclerotized tubercles. Endophallus without spines, aedeagus without terminal hook. **Hippobosca variegata** Megerle

Hippobosca camelina Leach, 1817
Figures 21–22, 29, 42, 45, 48, 50, 54

Hippobosca camelina Leach, 1817. *On the Genera and Species of Eproboscideous Insects*, p. 10.
Hippobosca bactriana Rondani, 1878. *Annali Mus. Civ. Stor. Nat. Giacomo Doria*, 12 : 165.
Hippobosca dromedarina Speiser, 1902. *Z. syst. Hymenopt. Dipterol.*, 2 : 176.
Hippobosca camelina Leach. Theodor & Oldroyd (1964) in Lindner, 65, Hippoboscidae, p. 26.

Head + thorax 6–7 mm. Wing length 9–10 mm. Sexual dimorphism distinct.
Head: Vertex about as wide as an eye, slightly narrower posteriorly. Parafrontalia broad in the middle, pointed anteriorly and posteriorly. Mediovertex narrower in the middle than a parafrontal. A row of short, thin orbital setae

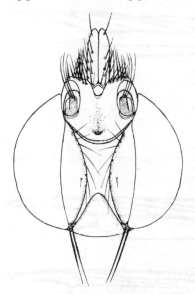

Fig. 42: *Hippobosca camelina* Leach. Head, dorsal

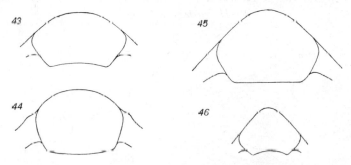

Figs. 43–46: Praesternum. 43. *Hippobosca equina* Linnaeus;
44. *H. variegata* Megerle; 45. *H. camelina* Leach;
46. *H. longipennis* Fabricius

to near the triangular postvertex. Two to three vertical setae at each side. Frontal horns broad, truncate, with light setae; indentation between horns rounded. Palps as long as the frons.

Thorax: With the typical white pattern of the genus. Prothorax visible dorsally, with a row of short setae and a white spot in the middle. Scutellum rhomboidal in the female, with a distinct posterior angle and with a row of 4–8 setae near the posterior margin. Scutellum shorter, rounded posteriorly, with a double row of 10–20 light setae in the male. Postnotal calli brownish, with a vertical row of setae. Praesternum triangular. Sternal plate yellowish brown, with dark spots on basisternum 2. Wings: R_1 and R_2+_3 end at the same point in the

25

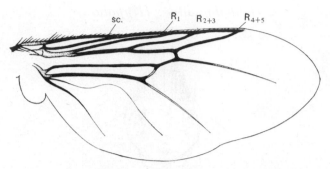

Fig. 47: *Hippobosca equina* Linnaeus. Wing

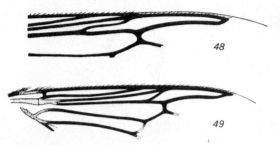

Figs. 48–49: Ending of R_1 and $R_4 + _5$ in the costa. 48. *H. camelina* Leach;
49. *H. variegata* Megerle

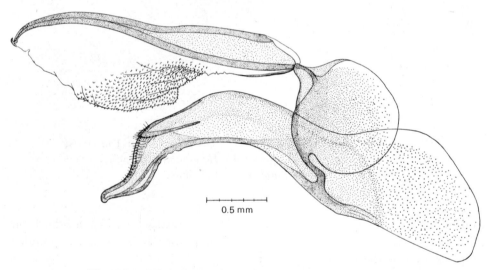

Fig. 50: *Hippobosca camelina* Leach. Male genitalia

costa or close together. R_2+_3 about half as long as the part of R_4+_5 distal to
r–m. Hairs on tegula and the thick basal part of the costa light. *Legs*: Claws
symmetrical. Pulvilli rudimentary. Empodium practically bare.
Male Abdomen: Tergite 1 + 2 with nearly straight posterior margin, with a
notch in the middle. Tergal plate 3 broad. Tergal plate 4 absent. Tergite 5 with

26

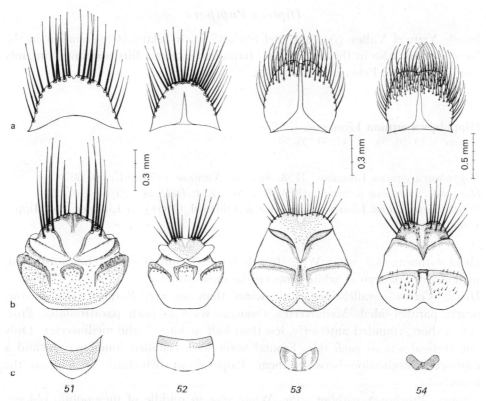

Figs. 51–54:   Anal and genital sclerites of female. 51. *Hippobosca equina*
Linnaeus; 52. *H. longipennis* Fabricius; 53. *H. variegata* Megerle;
54. *H. camelina* Leach
a – dorsal part of anal frame; b – genital sclerite; c – ventral genital plate

two small, elliptical lateral sclerites. Lateral sclerites of tergite 6 very large,
hemispherical, situated close together. Sclerites of tergite 7 small, elliptical,
situated close to the base of the genitalia. Sternite 1 trapezoidal. The whole abdomen covered with light setae. *Genitalia*: Aedeagus 2 mm long, straight at the
base, curved apically, with a small hook at the end. Dorsal ridges fused at the
apex. Endophallus large, with numerous small, triangular spines in the basal
half. Postgonites markedly curved, with an inner process at the dorsal margin.
Praegonites large, rounded, with numerous setae. Anal sclerite very long and
slender, triangular, with short setae at the end.
Female Abdomen: Tergal plate 3 less wide than in the male. Tergal plates 4
and 5 absent. Sclerites of tergite 6 more widely separated and smaller than in
the male, elliptical, situated close to the anus. Lateral sclerites of tergite 7
smaller than those of tergite 6. Dorsal part of anal frame trapezoidal, longer
than wide, with numerous setae. Genital sclerite as in Fig. 54.
Hosts: *Camelus bactrianus* L. and *C. dromedarius* L. Occasionally on horses;
bites man occasionally.
Distribution: Crete; S.W. Asia, mainly in deserts; N. Africa; to Kenya and
Nigeria in the south.

Israel: Yizre'el Valley (5) and Bet She'an (7) to Elat (16), common in the Negev, occurs also in the mountains (Jerusalem); about fifty specimens, mainly from the Negev; February to October.

Hippobosca equina Linnaeus, 1758
Figures 5, 19–20, 38, 43, 47, 51, 55–56

Hippobosca equina Linnaeus, 1758. *Systema Naturae* (10th ed.), p. 607.
Hippobosca taurina Rondani, 1879. *Boll. Soc. Ent. Ital.*, 11 : 25.
Hippobosca equina Linnaeus. Theodor & Oldroyd (1964) in Lindner, 65, Hippoboscidae, p. 28.

Head + thorax 4.5 mm. Wing length 6.6–8.0 mm. Colour chestnut-brown, with distinct pattern of white spots on the mesonotum.

Head: Vertex parallel-sided, little wider than an eye. Parafrontalia narrow, nearly parallel-sided. Mediovertex about as wide as both parafrontalia. Postvertex short, rounded anteriorly, less than half as long as the mediovertex. Only one vertical seta at each side. Frontal horns with rounded inner margin and a rounded invagination between them. Palps about two-thirds as long as the frons.

Thorax: Prothorax without setae. White spot in middle of mesonotum oblong-triangular. Scutellum short, rounded posteriorly, white in the middle half or slightly more. Five to eleven, usually eight, setae near the posterior margin. Postnotal calli white, inner side darker, with a vertical row of setae. Praesternum rounded anteriorly. Sternal plate whitish, with large dark spots on basisternum 2.

Wings: $R_2 + _3$ about as long as $R_4 + _5$ distal to r–m, ending markedly distal to R_1 in the costa. Section of costa between $R_2 + _3$ and $R_4 + _5$ about two to three times longer than section between R_1 and $R_2 + _3$. Veins darkly pigmented. Tegula and thick basal part of costa mainly with dark setae. *Legs*: Claws and pulvilli asymmetrical, the larger pulvillus at the shorter claw. Empodium practically bare.

Male Abdomen: Posterior margin of tergite 1 + 2 with a distinct notch in the middle. Tergal plates 3–5 well developed, 3 and 4 about half as wide as the abdomen, tergal plate 5 wider and broader laterally. Sclerites of tergite 6 small, rounded; sclerites of tergite 7 absent. *Genitalia*: Aedeagus short, 1 mm long, curved at the base, with two dorsal ridges which are not connected at the apex. Endophallus large, covered with spines. Postgonites triangular, with truncate end. Praegonites broadly rounded. Anal sclerite short, parallel-sided, with long setae at the end.

Female Abdomen: Tergal plates 3–5 smaller than in the male, triangular or elliptical. Sclerites of tergite 6 small, elliptical; sclerites of tergite 7 larger, situated close to the anus. Genital sclerite as in Fig. 51.

Hosts: Domestic horses and cattle, rarely other hosts.

28

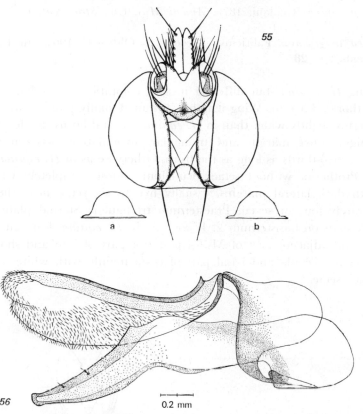

Figs. 55–56: *Hippobosca equina* Linnaeus. 55. head, dorsal (a, b – variation
of form of postvertex); 56. male genitalia

Distribution: Whole Palaearctic region south of Sweden, to Amboina in the
Oriental region; introduced in Pacific Islands to New Caledonia and Fiji; absent
in Africa south of the Sahara and in Australia, in spite of accidental introductions.
Israel: Whole country, rare in the Negev; numerous specimens examined; March
to October.

Hippobosca longipennis Fabricius, 1805
Figures 25–28, 30, 40–41, 46, 52, 57–58

Hippobosca longipennis Fabricius, 1805. *Systema Antliatorum*, p. 338.
Hippobosca capensis Olfers, 1816. *De Vegetativis et Animatis Corporibus . . .*, I,
p. 101.
Hippobosca francilloni Leach, 1817. *On the Genera and Species of Eproboscideous
Insects*, p. 8.
Hippobosca orientalis Macquart, 1842. *Mém. Soc. Sci. Agric. Lille*, p. 432.
Ornithomyia chinensis Giglioli, 1864. *Q. Jl Microsc. Sci.*, 4 : 23.

Hippobosca canina Rondani, 1878. *Annali Mus. Civ. Stor. Nat. Giacomo Doria,* 12 : 164.
Hippobosca longipennis Fabricius. Theodor & Oldroyd (1964) in Lindner, 65, Hippoboscidae, p. 28.

Resembling *H. equina,* but differing in size, coloration, genitalia, etc.
Head + thorax 3.5 mm. Wing length 5–6 mm. Usually pale brown.
Head: Vertex slightly wider than in *H. equina.* Frontal horns slender, triangular, with straight inner margin and triangular invagination between the horns. Palps about two-thirds as long as the frons. Otherwise as in *H. equina.*
Thorax: Prothorax without setae. Scutellum almost completely white, sometimes with dark lateral margins. Usually five to six setae near the posterior margin, rarely four or seven. Praesternum triangular. Sternal plate brownish, with dark spots on basisternum 2. *Wings:* As in *H. equina,* but veins light, except r–m and adjacent part of M_1+_2, posterior part of i–m and short parts of adjacent veins. Tegula and basal part of costa mainly with white setae, rarely a few black setae.

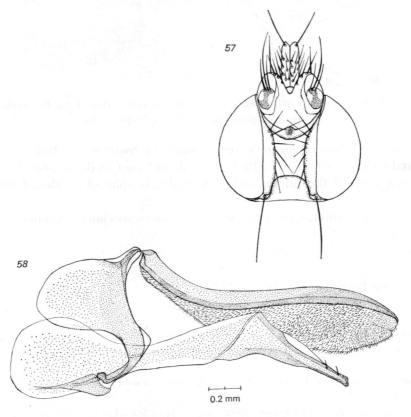

Figs. 57–58: *H. longipennis* Fabricius. 57. head, dorsal; 58. male genitalia

Male Abdomen: As in *H. equina*, but tergal plate 5 wider and not broader laterally, its anterior margin sometimes obtuse angular. *Genitalia*: Aedeagus distinctly longer than in *H. equina*, 1.6 mm, slightly S-shaped; dorsal ridges not connected at the apex. Endophallus as in *H. equina*. Postgonites more slender than in *H. equina*. Anal sclerite oblong-triangular, with 10–12 setae at the end. *Female Abdomen:* As in *H. equina*. Genital sclerite as in Fig. 52, much smaller than in *H. equina*.

Hosts: Mainly dogs in the Palaearctic region; wild carnivores in Africa, including Felidae; not yet found on the domestic cat.

Distribution: S. Europe and Mediterranean to China; rare in C. Europe, India, Africa.

Israel: Common throughout the country to Elat (16) and S. Sinai (22, 23); numerous specimens from domestic dog, *Meles meles* L., *Felis silvestris* Schreber (=*F. libyca* Forster), *Canis aureus* L., *Vulpes vulpes* L.; February to October.

Hippobosca variegata Megerle, 1803
Figures 44, 49, 53, 59–61

Hippobosca variegata Megerle, 1803. *Append. Catal. Insect.*, Nov. 1802, Vienna, p. 14.
Hippobosca maculata Leach, 1817. *On the Genera and Species of Eptobuscideous Insects*, p. 7.
Hippobosca bipartita Macquart, 1842. *Mém. Soc. Sci. Agric. Lille*, p. 432.
Hippobosca aegyptiaca Macquart, 1842. *Ibid.*, p. 431.
Hippobosca fossulata Macquart, 1842. *Ibid.*, p. 433.
Hippobosca sudanica Bigot, 1884. *Annls Soc. Ent. Fr.*, Ser. 6, Vol. 4, p. 69.
Hippobosca calopsis Bigot, 1885. *Ibid.*, 5:236.
Hippobosca aegyptiaca var. *bengalensis* Ormerod, 1895. *Vet. Rec.*, 8:82.
Hippobosca variegata Mergele. Theodor & Oldford (1964) in Lindner, 65, Hippoboscidae, p. 29.

Head+thorax 5.5 mm. Wing length 7–8 mm. Colour reddish brown to blackish brown, with sharply defined white spots on the mesonotum.
Head: Vertex slightly wider than an eye anteriorly, half as wide posteriorly. Parafrontalia broader in the middle. Mediovertex narrower than one parafrontal in the middle. Postvertex longer than wide at the base, pointed anteriorly. One vertical seta at each side. Frontal horns broad, truncate anteriorly, invagination between horns forming an angle of about 60°. Palps as long as the frons.
Thorax: Prothorax with a white spot in the middle and with a row of short setae. White spots on mesonotum sharply defined; spot in the middle transversely rectangular with a triangular anterior process, spot behind the transverse suture resembling the silhouette of a flying bird. Scutellum with rounded posterior margin and with a double, dense row of about twenty light setae, with three light spots on dark ground, the middle spot usually larger, triangular. Postnotal

31

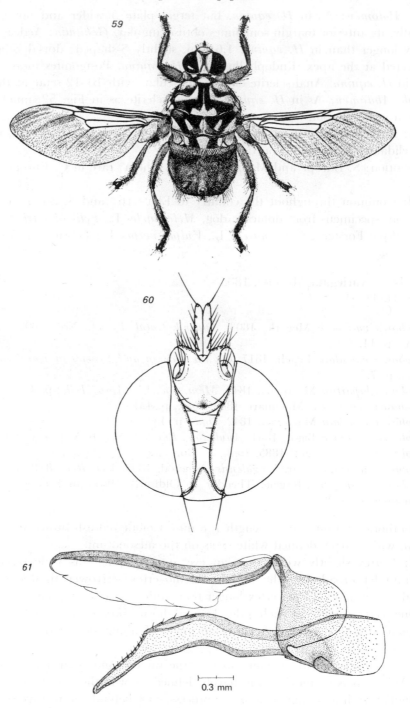

Figs. 59–61: *Hippobosca variegata* Megerle. 59. general view (after Byam and Archibald, 1921); 60. head, dorsal; 61. male genitalia

calli divided, with an inner ridge with a vertical row of long setae and with a shorter outer process with an irregular row of setae. Praesternum rounded-rectangular. Sternal plate brown or light brown with dark spots on basisternum 2. *Wings*: R_1 and R_2+_3 end at the same point in the costa or close together. R_2+_3 very short, little longer than r–m. Basis of radius sector with a row of 6–10 light spines. Section of costa between R_1 and R_4+_5 as long as part of R_4+_5 distal to r–m. *Legs*: Claws and pulvilli asymmetrical, the larger pulvillus at the shorter claw on fore- and mid-legs. Both pulvilli reduced on hind legs. Empodium practically bare.

Male Abdomen: Posterior margin of tergite 1+2 deeply incised. Tergal plates 3–5 absent. Sclerites of tergite 6 very large, rounded-triangular, resembling those of *H. camelina*; those of tergite 7 smaller, situated close to the base of the genitalia. *Genitalia*: Aedeagus resembling that of *H. camelina*, 1.8 mm long, but less curved and without terminal hook. Endophallus without spines. Postgonites slender, triangular, with rounded apex and a dorsal angle. Anal sclerite slender, triangular, but widening again at the base, with short setae at the end.

Female Abdomen: Sclerites of tergite 6 smaller than in the male; sclerites of tergite 7 larger than in the male. Numerous short, light spines on large, rounded, sclerotized tubercles in the middle of the posterior part of the ventral side. Genital sclerite as in Fig. 53.

Hosts: Domestic horses and cattle; bites man occasionally.

Distribution: Mainly Africa south of the Sahara, reaching the Palaearctic region in Iran and N. Africa; accidentally introduced in C. Europe (Czechoslovakia); Oriental region.

Israel: Not recorded, but may occur occasionally.

Tribe ORNITHOMYIINI Bequaert, 1954. *Entomologica Am.*, 34 : 16, redefined by Theodor & Oldroyd (1964) in Lindner, 65, Hippoboscidae, p. 18.

Genus O R N I T H O I C A Rondani, 1878

Annali Mus. Civ. Stor. Nat. Giacomo Doria, 12 : 159

Type Species: *Ornithoica beccariina* Rondani, 1878.

The genus has a number of primitive characters which distinguish it from the other genera of the tribe.

Small species, wing length 2.0–4.5 mm. Thorax usually dark, with light humeral calli and mesopleura.

Head: Rounded, broader than long, less dorso-ventrally flattened than in other genera of the tribe. Eyes large, nearly hemispherical, reaching to the posterior

33

margin. Parafrontalia very narrow, parallel-sided. Postvertex triangular, with rounded apex, with setae or bare. Three ocelli well developed. Vertical setae situated on a short process. Anterior frons reduced to a small triangle between the antennae. Antennae with rounded dorsal process, contiguous in the middle, situated in a common pit. Basal segment of antenna (scape) distinct, with setae. Arista spatula-shaped. Palps short and broad. Posterior ventral process broadly rounded with a row of short spines.

Thorax: Prothorax very narrow. Humeral calli rounded, not projecting, with spines laterally. Anterior spiracle rounded-oval, directed laterally. Mesopleuron visible dorsally, also with spines, which are more numerous in the female. Median longitudinal suture absent. Transverse suture interrupted in the middle. Scutellum rhomboidal, with four long setae. Postnotal calli rounded, with a vertical row of setae. Prosternum not divided, trapezoidal. Sternal plate with short setae and spines. *Wings*: Three cross veins. Costa interrupted between Sc and R_1. R_1 and R_4+_5 with short setae. R_4+_5 curved towards the costa in the basal half and adjacent to the costa in the apical half. Membrane with microtrichia, the distribution of which varies in the different species. *Legs*: Claws simple. Pulvilli large. Empodium feathered.

Abdomen: Tergite 1+2 short. Tergal plates 3–6 large and broad in both sexes, of about equal size in the female; tergal plate 6 of male smaller, sometimes divided. Tergite 7 of female with lateral sclerites close to the anus; sclerites absent in the male. Sternite 1 small, rounded. Small sclerites and tubercles with spines on the ventral side before the genital opening of the female, their arrangement and number varying in the different species. Curved spines on tubercles on the pleurae of the females of some species which are absent in the male.

Genitalia: Aedeagus slightly curved, sclerotized laterally and dorsally, with a large endophallus. Postgonites triangular. Praegonites very small. Genital sclerite of female with plates that cover the genital opening and with a triangular plate with spines inside the genital opening. Ventral genital plate of varying form, T-shaped, triangular, etc.

Twenty-two species which are distributed world-wide throughout the tropics and subtropics. One African species (*O. turdi*) occurs rarely in the Mediterranean and C. Europe. Hosts mainly Passeriformes, Strigiformes and Ciconiiformes, but also species of other orders.

Ornithoica turdi (Latreille, 1811)
Figures 62–72

Ornithomyia turdi Latreille, 1811. *Encyclopédie méthodique, Insectes,* VIII, p. 544.
Stenopteryx pygmaea Macquart, 1835. *Histoire naturelle des insectes,* p. 644.
Ornithoica turdi Latreille. Theodor & Oldroyd (1964) in Lindner, 65, Hippoboscidae, p. 31.

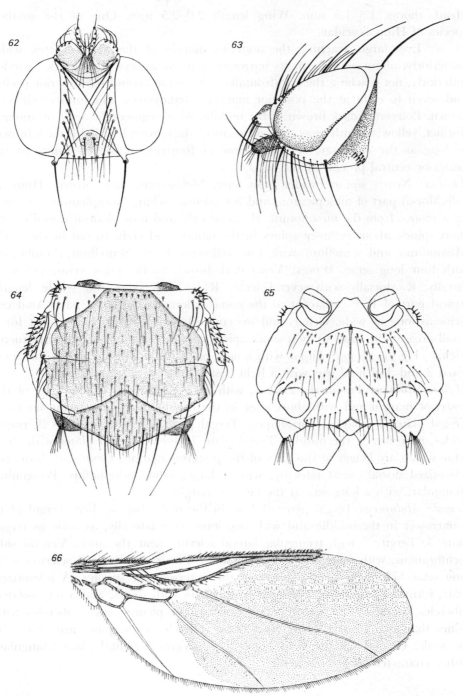

Figs. 62–66: *Ornithoica turdi* (Latreille). 62. head, dorsal; 63. head, lateral; 64. thorax, dorsal; 65. thorax, ventral; 66. wing

Head+thorax 1.3–1.5 mm. Wing length 2.0–2.5 mm. One of the smallest species of Hippoboscidae.

Head: Eyes large, reaching the posterior margin of the head. Vertex wider posteriorly, about as wide in its narrowest part as an eye. Postvertex rounded anteriorly, not reaching the parafrontalia. Two setae behind the anterior ocellus and seven to eight at the posterior margin. Mediovertex and lunula yellowish brown. Postvertex dark brown in the middle. Basal segment (scape) of antenna distinct, yellowish, with setae. Second segment dark, rounded. Palps dark brown, as long as the antennae. A double row of fourteen to sixteen spines at the posterior ventral process of the head.

Thorax: Nearly square, wider than long. Mesonotum dark brown. Humeral calli, dorsal part of mesopleuron, and notopleuron white. Notopleuron separated by a groove from the mesonotum. Humeral calli and mesopleuron dorsally with short spines, about eighteen spines in the female and eight to ten in the male. Mesonotum and scutellum with fine yellowish hairs. Scutellum rhomboidal, with four long setae. *Wings*: Veins dark brown to the cross veins, yellowish distally. R_1 dorsally with several setae, R_4+_5 with setae the whole length. Apical part of R_4+_5 adjacent to the costa as long as the basal part. Anal cell twice as long as wide. Cell r_5 and m_2 covered with microtrichia, except for a small area at the base. Only a small apical part of cell m_4 covered with microtrichia (Fig. 66). *Legs*: Tibiae with a light band in the middle and at the apex. Tarsi 2 and 3 of hind legs with a light basal band.

Male Abdomen: Tergite 1+2 short, with two indentations at the sides of the posterior margin and with short setae in the middle and longer setae laterally. Tergal plates 3–5 large, strip-shaped. Tergal plate 6 half as wide as 3, trapezoidal, narrower in the middle. Tergal plate 7 absent. Tergal plates with short setae which are longer at the sides of the posterior margin. *Genitalia*: Aedeagus sclerotized dorsally and laterally, with a large, bifid endophallus. Postgonites triangular, with a long seta at the ventral margin.

Female Abdomen: Tergal plates 3–5 as in the male, but smaller. Tergal plate 6 narrower in the middle and with two long setae laterally, as wide as tergal plate 3. Tergite 7 with triangular lateral sclerites near the anus. Ventral side membranous, with setae in the anterior half. Two pairs of small sclerites with long setae before the genital opening, the anterior pair smaller. A sclerotized plate, formed by fusion of five to seven tubercles with spines, and a few isolated tubercles lateral to the sclerites. Anterior part of pleurae with tubercles with spines that are directed posteriorly. These tubercles with spines are absent in the male. Genital sclerite as in Figs. 71–72. Ventral genital plate triangular. Other characters as for the genus.

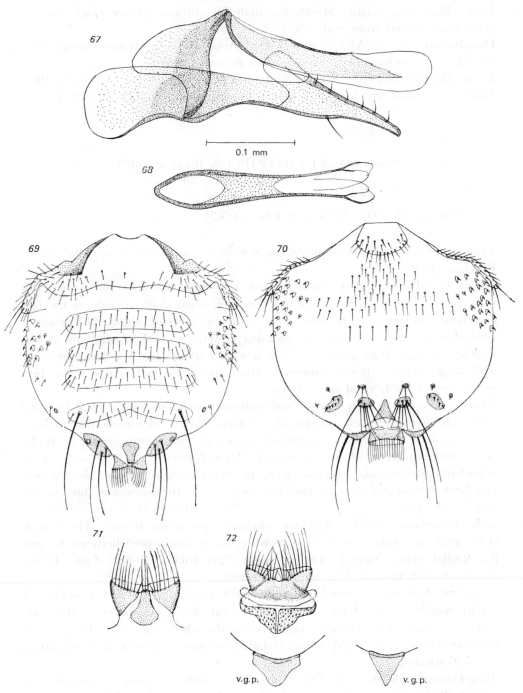

Figs. 67–72: *Ornithoica turdi* (Latreille). 67. male genitalia; 68. male, aedeagus, dorsal; 69. female abdomen, dorsal; 70. female abdomen, ventral; 71. female, anal frame, dorsal; 72. female, genital sclerite with ventral genital plates (v.g.p.)

Hosts: Rather unspecific. Mainly on small Passeriformes, more rarely on Falconiformes, Strigiformes and others.

Distribution: Mainly Africa; rare in C. Europe and the Mediterranean. Most records are from localities near migration routes of birds.

Israel: Bet Guvrin (10), two females from *Lanius senator* L.; 15 August 1967. Netanya (8), one female from *Streptopelia senegalensis* L.; 22 July 1972.

Genus O R N I T H O P H I L A Rondani, 1879
Boll. Soc. Ent. Ital., 11 : 20.

Ornitheza Speiser, 1902. *Természetr. Füz.*, 25 : 329.

Type Species: *Ornithomyia metallica* Schiner, 1864, as *Ornithophila vagans* Rondani, 1879.

Head: Broadly oval. Eyes large, nearly hemispherical, with small posterior orbits. Parafrontalia with curved inner margin, about half as wide as the mediovertex. Postvertex with three ocelli near the anterior angle. One pair of vertical setae. Lunula triangular, rounded posteriorly, with a small pit in the middle. Anterior frons small, Y-shaped, with curved lateral horns. Antennae with large, pointed, dorsal processes. Arista spatula-shaped. Palps short, projecting only little beyond the antenae.

Thorax: Anterior margin straight and with strongly projecting, conical humeral calli with long setae. Anterior spiracle on dorsal surface of mesonotum, at basis of humeral callus. Median longitudinal suture of mesonotum complete, transverse suture interrupted in the middle. Mesopleuron with long setae. One notopleural, one postalar and one or two praescutellar setae on each side. Scutellum broadly triangular, with four long setae near the apex and fine yellow hairs at the posterior margin. Postnotal calli rounded, with a vertical row of setae. Prosternum divided into two triangular processes. *Wings*: Three cross veins, anal cross vein close to base of wing. Costa interrupted between Sc and R_1. Radial veins situated close to costa. Microtrichia absent. *Legs*: Claws double. Pulvilli normal. Empodium feathered.

Abdomen: Abdomen of female with very small tergal plates 3–5. Tergites 6 and 7 with widely separated lateral sclerites. Tergal plates 3–4 larger in the male and tergal plates 5–6 extending nearly across the whole abdomen. Tergite 7 of male without sclerites. Basal sternite rounded. Sternite 5 of male with two large, elliptical sclerites.

Male Genitalia (known only for *O. metallica*): Aedeagus long, narrow, slightly curved. Postgonites slender, pointed. Praegonites finger-shaped.

Three species, of which one is widely distributed in the Palaearctic region and in the tropics and subtropics of the Old World. The other two species are known only from a few specimens in the Mediterranean.

38

Ornithophila metallica (Schiner, 1864)
Figures 34, 74–82

Ornithomyia metallica Schiner, 1864. *Fauna Austriaca*, II, p. 646.
Ornithophila vagans Rondani, 1879. *Boll. Soc. Ent. Ital.*, 11 : 21.
Ornitheza pallipes Speiser, 1904. *Z. syst. Hymenopt. Dipterol.*, 4 : 177.
Ornithophila metallica Schiner. Theodor & Oldroyd (1964) in Lindner, 65,
Hippoboscidae p. 34. (Full synonymy in Maa, 1963.)

Head + thorax 2.5 mm. Wing length 4–5 mm. Colour blackish brown.
Head: Parafrontalia widest behind the middle. One row of orbital setae in
the anterior half of the parafrontalia, one very long, black seta anteriorly and
a usually light, long seta in the middle. Area of ocelli on postvertex, mediovertex
and posterior part of parafrontalia dark brown, but coloration variable. Dorsal
process of antennae very large, pointed, with black setae. Palps very short, dark
brown.
Thorax: Mesonotum dark brown. Humeral calli conical, pointed, transparent,
whitish with dark basis, with two to three long setae which are directed posteriorly, and a few spines at the apex. Longitudinal and transverse sutures of
mesonotum distinct, transverse suture interrupted in the middle. Notopleuron
whitish, with a long seta. Other parts of mesonotum also sometimes whitish.
Mesopleuron whitish, with several black setae. A group of short, black setae
behind the postalar setae. Scutellum dark brown, with a light stripe at the base
and with four long setae at the apex. Sometimes a light spot at the apex. Postnotal calli rounded, with a vertical row of setae. *Wings*: R$_{2+3}$ nearly contiguous with the costa in its apical part, R$_{4+5}$ forming a very acute angle with
the costa. First basal cell twice as long as second. Cross vein i–m situated in the
middle between r–m and the anal cross vein. Anal cell very short. *Legs*: Claws
double, black, with a light, pointed, basal tubercle.

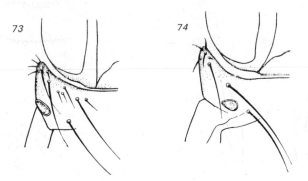

Figs. 73–74: Humeral callus and anterior spiracle. 73. *Ornithomyia
avicularia* (Linnaeus); 74. *Ornithophila metallica* (Schiner)
(on the dorsal surface)

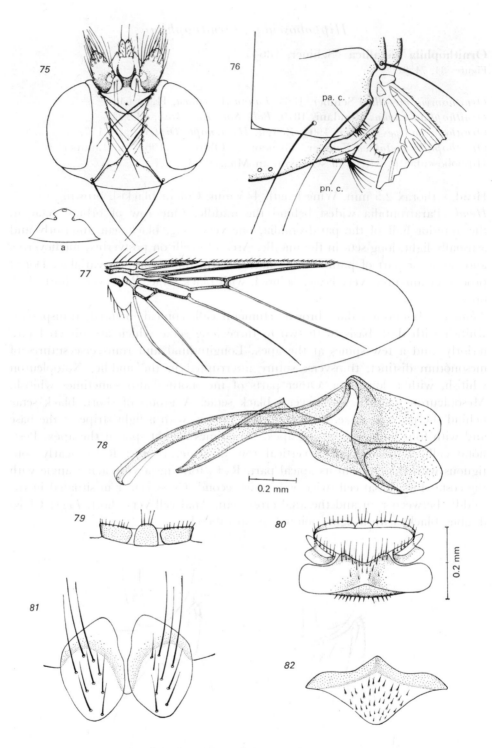

Figs. 75–82: *Ornithophila metallica* (Schiner). 75. head, dorsal; 75a. different
form of postvertex; 76. group of spines on the postalar callus; 77. wing;
78. male genitalia; 79. female, anal frame, dorsal; 80. genital sclerite;
81. ventral genital plates; 82. plate inside genital opening

pa.c. – postalar callus; pn.c. – postnotal callus

Male Abdomen: Tergite 1+2 with straight posterior margin, dark at the base, light laterally and posteriorly, with three to four rows of setae which are long at the lateral corners. Tergal plates 3–4 narrow, about as wide as the scutellum. Tergal plate 5 wide, curved, nearly as wide as the abdomen. Tergal plate 6 similar, but less wide. Tergal plate 7 absent. Sternite 1 rounded, with short setae posteriorly. Two large, elliptical scleritis on sternite 5 near the genitalia. *Genitalia*: Aedeagus 0.8 mm, slightly S-shaped, with a short ventral hook at the apex. Postgonites slender, triangular, pointed. Praegonites finger-shaped, with two to three long setae. Anal sclerite absent.

Female Abdomen: Tergal plates 3–5 very small. Tergite 6 with widely separated lateral sclerites with three to four setae. Tergite 7 with semi-circular lateral sclerites near the anus. Two large, angular sclerites with long setae before the genital opening on the ventral side. Abdomen covered with short black setae on small, sclerotized tubercles. Anal frame divided into three parts on the dorsal side. Genital sclerite as in Figs. 80–82, with plates which cover the genital opening and a triangular plate inside the opening, which is covered with spines.

Hosts: Very unspecific, recorded from sixty-three species of birds, mainly Passeriformes and Coraciiformes. Records from Falconiformes are probably mainly due to transition from prey.

Distribution: Southern parts of the Palaearctic region; Japan; tropics of the Old World.

Israel: Yizre'el Valley (5), Coastal Plain, Jerusalem; fourteen specimens from *Sturnus vulgaris* L., *Corvus cornix* L., *Passer domesticus* L.; April to September.

Genus O R N I T H O M Y I A Latreille, 1802

Histoire naturelle des Crustacées et des Insectes, III, p. 466.

Type Species: *Hippobosca avicularia* Linnaeus, 1758.

Head: Wider than long or as long as wide. Eyes large, a third or fourth as wide as the head. Parafrontalia narrow, with a row of long and short setae. Mediovertex long. Postvertex triangular, ocelli present or absent. One pair of vertical setae. Lunula large, well defined. Anterior frons narrow, with horns which are sclerotized only on the outside. Antennae with long dorsal process. Arista spatula-shaped. A row of long jugular setae, the posterior setae very long. A row of setae or spines at the posterior ventral process of the head.

Thorax: Prothorax narrow, wider laterally. Humeral calli long, conical, reaching beyond the posterior margin of the eyes. Anterior spiracle directed laterally.

41

Median longitudinal suture and transverse suture of mesonotum distinct, transverse suture interrupted in the middle. Notopleuron divided from mesonotum by a suture, with a long seta. Scutellum large, rhomboidal, with a row of setae near the posterior margin and on the surface in some species. Postnotal calli rounded, without setae. Prosternum divided into two blunt, triangular processes. *Wings*: Large, longer than the body, with three cross veins, cross vein i–m sometimes only partly pigmented. Costa interrupted before the end of Sc. Membrane covered with microtrichia to a varying extent. *Legs*: Claws double. Pulvilli normally developed. Empodium feathered.

Abdomen: Tergite 1+2 with nearly straight posterior margin. Tergal plates 3–5 large in the male, small in the female. Tergite 6 with lateral sclerites. Tergal plate 7 absent in both sexes. Sternite 1 small, trapezoidal. Ventral side otherwise membranous.

Genitalia: Aedeagus straight, only slightly tapering, with a small endophallus. Postgonites triangular, straight. Praegonites oblong, small, with a few setae at the end. Genital sclerite of female with plates that cover the genital opening and a second plate inside it.

The genus is distributed throughout the world. Some species (*avicularia* group) are rather unspecific, others (*biloba* group) occur only on Hirundinidae.

Four W. Palaearctic species, of which only one has been recorded from Israel. Other species may, however, be introduced by migrating birds.

Key to the West Palaearctic Species of Ornithomyia

1. Head wider than long. Anterior ocellus on a line connecting the posterior margin of the eyes. Eyes about one-third as wide as the head. Parafrontalia with a long seta anteriorly and one in the middle, other setae short. Scutellum with a row of long setae near the posterior margin. Wings with microtrichia only in cell r_5 and m_2, rarely a narrow stripe in cell m_4 (*avicularia* group) 2
   – Head as wide as long (from the base of the indentation between the frontal horns to the posterior margin). Parafrontalia with a uniform row of setae to the middle. Anterior ocellus situated distinctly posterior to the line connecting the posterior margin of the eyes. Vertex extending beyond the posterior margin of the eyes for the width of an eye. Width of eyes one-fourth of width of head. Scutellum with a row of six setae posteriorly and several long and short setae on the surface. Microtrichia covering cell r_5 and m_2 completely and also covering the apical part of cell m_4. Section of costa between R_1 and R_{2+3} nearly twice as long as section between R_{2+3} and R_{4+5}. **O. biloba** Dufour

2. Large species. Wing length 6–7 mm. Usually eight setae on scutellum. Section of costa between R_1 and R_{2+3} nearly twice as long as section between R_{2+3} and R_{4+5}. Cross vein i–m about four times as long as cross vein r–m.
   O. avicularia Linnaeus

– Smaller species. Wing length 3.5–5.5 mm. Section of costa between R_1 and $R_2 + _3$ as long as, or little longer than, section between $R_2 + _3$ and $R_4 + _5$. Cross vein i–m shorter   3

3. Wing length 3.5–4.5 mm. Colour light brown, usually with a light pattern on the mesonotum. Usually four setae on the scutellum. Two dark brown triangular spots on the ventral side of the head which do not reach the long jugular seta. Microtrichia in cell r_5 separated by broad, bare stripes from the veins and with a bare spot near the apical end of $R_4 + _5$.   **O. fringillina** Curtis
– Wing length 4.5–6.0 mm. Colour dark brown. Usually six setae on the scutellum. The triangular spots on the ventral side of the head reach to the long jugular setae. Microtrichia cover cell r_5 completely to the veins, except in a small basal area.   **O. chloropus** Bergroth

Ornithomyia avicularia (Linnaeus, 1758)
Figures 2–4, 7, 10–12, 15, 35, 73, 83, 89–90

Hippobosca avicularia Linnaeus, 1758. *Systema Naturae* (10th ed.), p. 607.
Hippobosca corvi Scopoli, 1763. *Entomologia Carniolica*, p. 377.
Ornithomyia viridula Meigen, 1830. *Systematische Beschreibung der Europäischen zweiflügligen Insekten*, VI, p. 233.
Ornithomyia nigricornis Erichson, 1842. *Arch. Naturgesch.*, 8 : 274.
Ornithomyia opposita Walker, 1849. *List of the Dipterous Insects in the British Museum*, IV, p. 1145.
Hippobosca oculata Motschulsky, 1859. *Bull. Soc. Nat. Moscou*, 32 : 504.
Ornithomyia avicularia Linnaeus. Theodor & Oldroyd (1964) in Lindner, 65, Hippoboscidae, p. 37.

Head+thorax 2.7–3.5 mm. Wing length 6.0–7.0 mm. Colour light brown, without dark spots on the ventral side of the head and thorax.
Head: Wider than long, posterior margin slightly curved, without angle near the vertical setae. Eyes large, about a third as wide as the head. Mediovertex half as long as head. Postvertex triangular, dark anteriorly. Palps as long as the lunula. A row of four to five spines at the posterior ventral margin of the head.
Thorax: Sternal plate of thorax without dark spots. Scutellum large, rhomboidal, with a row of eight setae posteriorly, rarely seven or nine. *Wings*: Cross vein i–m about four times as long as r–m, unpigmented in its greater part. Section of costa between R_1 and $R_2 + _3$ twice as long as section between $R_2 + _3$ and $R_4 + _5$. Microtrichia only in middle of cell r_5, leaving broad stripes near the veins and an area near the apical end of $R_4 + _5$ bare. Two to three narrow stripes of microtrichia in cell m_2.
Male Abdomen: Tergal plates 3 and 4 about half as wide as the abdomen. Tergal plate 5 wider, curved, with long setae laterally. Tergite 6 with oblique, elliptical, lateral sclerites with three to four setae. Sternite 1 small, rounded, with several rows of short black spines in the posterior half. *Genitalia*: Aedeagus 1.2 mm long, straight, slightly tapering. Anal sclerite strip-like, without setae.

43

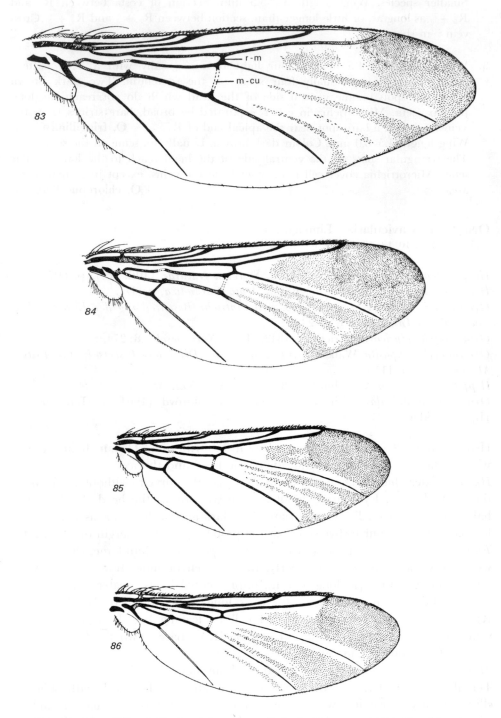

Figs. 83–86: Wings. 83. *Ornithomyia avicularia* (Linnaeus); 84. *O. chloropus* Bergroth; 85–86. *O. fringillina* Curtis

Female Abdomen: Tergal plates 3–5 small, triangular or elliptical. Lateral sclerites of tergite 6 larger than in the male, usually with three to four long setae. Two to three rows of setae before the genital opening and two long setae and a group of spines lateral to the anus. Genital sclerite as in Fig. 90, with plates that cover the genital opening.

Hosts: Mainly large Passeriformes, Strigiformes, Falconiformes and others.

Distribution: Whole Palaearctic region to 57° lat. N.; tropics of the Old World, to New Zealand in the South.

Israel: Not recorded, but may be introduced by migrating birds.

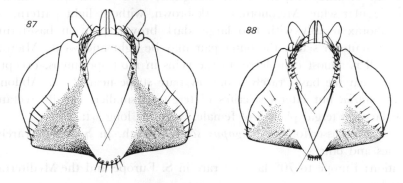

Figs. 87–88: Head, ventral. 87. *Ornithomyia chloropus* Bergroth; 88. *O. fringillina* Curtis

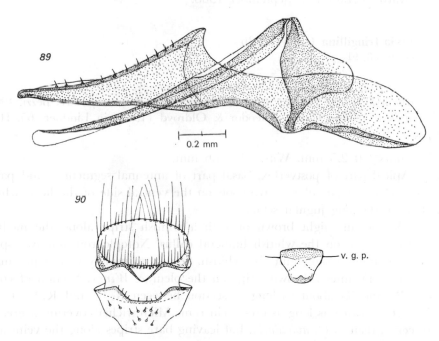

Figs. 89–90: *Ornithomyia avicularia* (Linnaeus). 89. male genitalia; 90. female, genital sclerite and ventral genital plate (v.g.p.)

Ornithomyia chloropus Bergroth, 1901
Figures 84, 87

Ornithomyia chloropus Bergroth, 1901. *Meddn Soc. Fauna Flora Fenn.*, 27 : 146.
Ornithomyia lagopodis Sharp, 1907 *Entomologist's Mon. Mag.*, 54 : 59.
Ornithomyia fringillina Curtis, 1836. Bequaert, 1954 (pro parte).
Ornithomyia chloropus Bergroth. Theodor & Oldroyd (1964) in Lindner, 65, Hippoboscidae, p. 38.

Head+thorax 2.5–3.0 mm. Wing length 4.5–5.5 mm. Colour dark brown.
Ventral side of head with two long, triangular, dark brown spots which reach the long jugular setae. Mesonotum dark brown, without light pattern. Sternal plate of thorax usually with two large, dark brown spots on basisternum 2. Usually six scutellar setae; the outer pair may be light and weak. Microtrichia covering cell r_5 almost completely to the veins in most specimens, except for a small area near the base; rarely a narrow bare stripe near R_4+_5. Abdomen as in *O. avicularia*, but lateral sclerites of tergite 6 smaller, usually rectangular, about as large as tergal plate 5 of female, with one long seta.
Hosts: Mainly Passeriformes, *Lagopus scoticus* Lath. in Scotland, rarely Falconiformes and others.
Distribution: Europe to 70° lat. N.; rare in S. Europe and the Mediterranean; Siberia; Japan.
Israel: Recorded only once, Petah Tiqwa (8); from *Crex crex* L., apparently on southward migration; 9 September 1960.

Ornithomyia fringillina Curtis, 1836
Figures 85–86, 88, 91

Ornithomyia fringillina Curtis, 1836. *British Entomology*, VIII, Pl. 585.
nec *Ornithomyia tenella* Schiner, 1964. Rondani, 1879; Bergroth, 1901; Bezzi, 1905.
Ornithomyia fringillina Curtis. Theodor & Oldroyd (1964) in Lindner, 65, Hippoboscidae, p. 39.

Head+thorax 2.0–2.5 mm. Wings 3.5–4.6 mm.
Head: Apical part of postvertex, basal part of antennal segment 2 and palps dark brown. Two triangular brown spots on the ventral side of the head which do not reach the long jugular setae.
Thorax: Mesonotum light brown or with a whitish stripe along the median suture, a dark spot on the whitish humeral callus. Notopleuron and two spots at anterior margin of scutellum also whitish. Scutellum usually with 4 praemarginal setae. Sometimes a brown stripe on the pleurae. *Wings*: Section of costa between R_1 and R_2 about as long as section between R_2+_3 and R_4+_5. Cross vein i–m about twice as long as cross vein r–m. Microtrichia covering a greater part of cell r_5 than in *O. avicularia*, but leaving bare stripes along the veins and a bare area near the end of R_4+_5 Area of microtrichia on ventral side elliptical. *Legs*: Basitarsus of hind legs often dark.

46

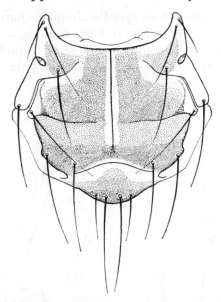

Fig. 91:   *Ornithomyia fringillina* Curtis. Thorax, dorsal

Abdomen: Arrangement of tergal plates and chaetotaxy as in *O. avicularia*, but setae shorter and thinner. Tergal plate 5 of female only slightly wider than tergal plate 4 and with only one or two setae at each side of the posterior margin. Lateral sclerites of tergite 6 of male relatively smaller and usually with one or two long setae.

Male Genitalia: As in *O. avicularia,* but more slender. Length of aedeagus 0.55–0.65 mm.

Hosts: Mainly small Passeriformes.

Distribution: N. Palaearctic to about 60° lat. N.; introduced in Australia and New Zealand.

Israel: Not recorded, but likely to be introduced by migrating birds.

Ornithomyia biloba Dufour, 1827
Figures 92–96

Ornithomyia biloba Dufour, 1827. *Annls Sci. Nat.,* 10:243.
Ornithomyia ptenoletis Loew, 1857. *Wiener Ent. Mschr.,* 1:9.
Ornithomyia tenella Schiner, 1864. *Fauna Austriaca,* II, p. 646.
Ornithomyia biloba Dufour. Theodor & Oldroyd (1964) in Lindner, 65, Hippoboscidae, p. 39.

Head+thorax 2.5 mm. Wing length 5–6 mm.
Head: Nearly round, as long as wide. Eyes small, about a fourth of the width of the head, with distinct posterior inner angle. Anterior frons V-shaped, narrower than in *O. avicularia*. The vertex extends behind the posterior margin of

the eyes by about the width of an eye. Parafrontalia narrow anteriorly, nearly twice as wide posteriorly. Orbital setae long, forming a complete row to the middle of the mediovertex. Anterior ocellus situated behind the posterior margin of the eyes. Palps as long as the antennae. A row of six to eight spines at the posterior ventral margin of the head.

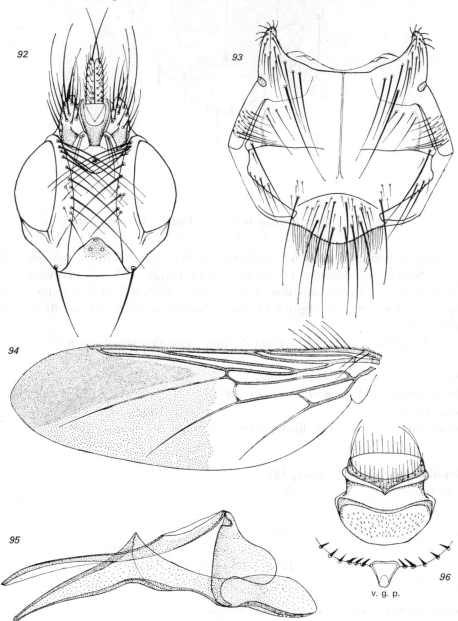

Figs. 92–96: *Ornithomyia biloba* Dufour. 92. head, dorsal; 93. thorax, dorsal; 94. wing; 95. male genitalia; 96. female, genital sclerite, with ventral genital plate (v.g.p.)

Thorax: Setae on mesonotum more numerous than in *O. avicularia*, particularly on the humeral calli. Scutellum usually with six setae posteriorly, and several long and short setae on the surface. Two to three postalar setae on each side. *Wings*: Venation resembling that of *O. avicularia*, but cross vein i–m distinctly shorter, two and a half times as long as r–m. The microtrichia cover cells r_5 and m_2 completely and also the apical part of cell m_4, so that the whole apical two-thirds of the wing are covered with microtrichia.

Male Abdomen: Tergal plates 3 and 4 about a third or fourth as wide as the abdomen, tergal plate 5 wider, about half as wide as the abdomen, slightly curved. Tergite 6 with oblique, triangular, lateral sclerites with three to four long setae. Sternite 1 with rounded posterior margin and with three to four rows of short setae in the posterior half. Pleurae only with setae. *Genitalia*: Aedeagus slender, slightly S-shaped, 1 mm long. Postgonites triangular, slightly curved. Praegonites small, finger-shaped, with two to three setae.

Female Abdomen: Tergal plates 3–5 small, triangular or elliptical. Lateral sclerites of tergite 6 as in the male. Pleurae anteriorly with a group of spines on sclerotized tubercles which are absent in the male. A row of spines and one or two rows of setae at the anterior margin of the genital opening. Genital sclerite as in Fig. 96. Numerous long spines on sclerotized tubercles at the end of the abdomen.

Hosts: Hirundinidae, mainly *Hirundo rustica* L.

Distribution: W. Palaearctic region; Africa.

Israel: Not recorded, but likely to be introduced by migrating swallows.

Genus C R A T A E R I N A Olfers, 1816
De Vegetativis et Animatis Corporibus..., I, p. 101.

Oxypterum Leach, 1817. *On the Genera and Species of Eproboscideous Insects*, I, p. 3.
Anapera Meigen, 1830. *Systematische Beschreibung der Europäischen zweiflügligen Insekten*, VI, p. 234.
Chelidomyia Rondani, 1879. *Boll. Soc. Ent. Ital.*, 11:15.

Type Species: *Ornithomyia pallida* Latreille, 1812, as *Crataerina lonchoptera* Olfers, 1816.

Head: As wide as long or markedly longer than wide. Posterior part of head prolonged in a plate. Eyes small, with distinct anterior and posterior inner angles, about a third as long and a fifth as wide as the head. Parafrontalia and posterior orbits very wide, with a row of orbital setae at the inner margin of the parafrontalia in the anterior half. Postvertex rounded or triangular. Ocelli absent. Anterior frons not clearly separated from lunula, with curved horns which are sclerotized only on the outer side. Antennae with long dorsal process. Arista spatula-shaped. One or two vertical setae at each side. A row of short setae at the posterior ventral process.

Thorax: Anterior margin deeply concave. Humeral calli very large, conical, with blunt apex, with several setae. Median longitudinal suture distinct, transverse suture interrupted in the middle. Notopleuron small, with one seta. Scutellum with curved anterior margin and nearly straight posterior margin, with a posterior row of setae and several setae on the surface. Middle of mesonotum and pleurae bare. *Wings*: Narrow, lanceolate or oval, non-functional. Venation complete, veins concentrated near costa. Costa distinct to end of R_4+_5. M_1+_2 and M_3+_4 indistinct distal to cross veins. Three cross veins. Venation very variable. *Legs*: Claws very long, double. Pulvilli normal. Empodium feathered. *Abdomen*: Very wide in older specimens. Tergal plates 3 and 4 of male large or small, sometimes absent, tergal plate 5 always wide. Tergal plates 3 and 4 absent, tergal plate 5 small or absent in female. Tergite 6 with lateral sclerites. Tergal plate 7 absent in both sexes.

Male Genitalia: Resembling those of *Ornithomyia*. Aedeagus straight. Postgonites triangular, very long.

Five species, of which one is Palaearctic, two occur in Africa and reach the south of the Palaearctic region, one is Oriental and one American.

Crataerina melbae (Rondani, 1879)
Figures 97–102

Chelidomyia melbae Rondani, 1879. *Boll. Soc. Ent. Ital.*, 11: 17.
Crataerina longipennis Austen, 1926. *Parasitology*, 18: 357.
Crataerina propinqua Austen, 1926. *Ibid.*, p. 358.
Crataerina melbae Rondani. Theodor & Oldroyd (1964) in Lindner, 65, Hippoboscidae, p. 43.

Head + thorax 4 mm. Wing length 7–8 mm.
Head: Oblong-oval. Posterior orbits as long as an eye. Parafrontalia much wider posteriorly. Mediovertex narrowest in the middle, markedly wider anteriorly. Postvertex narrow, rounded anteriorly. Clypeus distinct. Palps as long as the frons.
Thorax: Markedly concave anteriorly, twice as wide as long in the middle. Humeral calli very large, conical. *Wings*: Reaching well beyond the end of the abdomen, four times longer than wide, with a long, narrow apical part, which is about a third as long as the wing. Form of wings very variable, posterior margin sometimes concave, sometimes convex. Cross vein r–m situated in about the middle of the costa. *Legs*: Long and strong, hind femur 3 mm long.
Male Abdomen: Tergal plates 3 and 4 about a third as wide as the abdomen, tergal plate 5 nearly as wide as the abdomen, with long setae laterally at the posterior margin. Tergal plate 6 less wide, more curved and frequently divided into elliptical lateral sclerites, which are narrowly connected in the middle in some specimens. *Genitalia*: Aedeagus 1.35 mm long, slightly tapering. Postgonites triangular, slender. Anal sclerite strip-like, broader posteriorly, without setae.

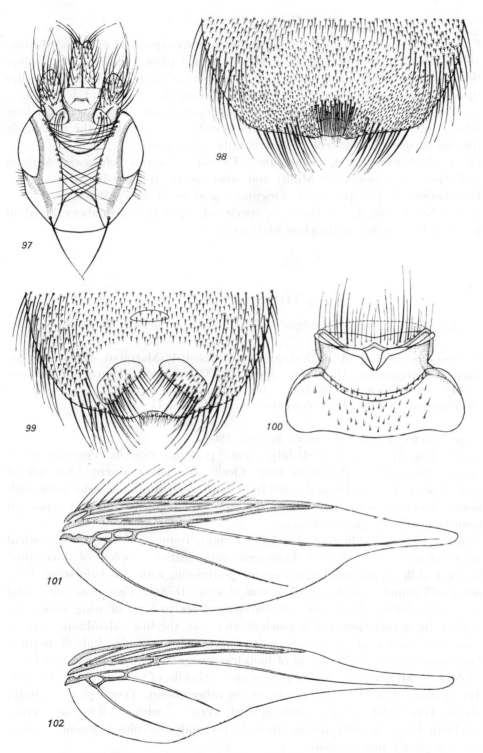

Figs. 97–102: *Crataerina melbae* (Rondani). 97. head, dorsal;
98. female abdomen, ventral, posterior part; 99. female abdomen, dorsal,
posterior part; 100. female, genital sclerite; 101. wing;
102. wing (*propinqua* Austen)

Female Abdomen: Broader than in the male, transverse-oval or rounded-triangular. Tergal plate 3 very small or absent. Tergal plate 4 slightly larger, elliptical. Tergal plate 5 larger than 4, strip-like. Tergal plate 6 divided into elliptical, oblique, lateral sclerites. A bare area before the sclerites of tergite 6. A dense group of long setae before the genital opening and large groups of long setae at the end of the abdomen. Genital sclerite as in Fig. 100, the part inside the genital opening markedly wider than the outer part.

Hosts: *Apus apus* L., *A. melba tuneti* Tschusi, *A. affinis galilejensis* Antinori and other. *A. aequatorialis* Müller and other species of *Apus* in Africa.

Distribution: S. Europe; Syria; Oriental region to Malaya; Africa.

Israel: Not recorded, but the three species of *Apus* recorded above breed in Israel and C. *melbae* is therefore likely to occur.

Genus I C O S T A Speiser, 1905

(=*Lynchia* auct., nec Weyenbergh, 1881) *Z. syst. Hymenopt. Dipterol.*, 5 : 358.

Olfersia Say, 1823. *J. Acad. Nat. Sci. Philad.*, 3 : 101; Macquart (1835) *Histoire naturelle des insectes*, II, p. 640; Speiser (1902) *Z. syst. Hymenopt. Dipterol.*, 2 : 149 (nec Wiedemann, 1830).
Ornithoponus Aldrich, 1923. *Insecutor Inscit. Menstr.*, 11 : 77.

Type Species: *Icosta dioxyrhina* Speiser, 1904.

Head: Broadly oval, with slightly curved posterior margin. Posterior orbits distinct, but narrow. Postvertex large. Ocelli reduced or absent. One pair of vertical setae. Lunula large, divided from frons by a suture. Anterior frons with horns. Antennae with short dorsal process, basal segment distinctly separated from frons. Arista spatula-shaped.

Thorax: Humeral calli conical, relatively short, rounded. Median longitudinal suture of mesonotum distinct. Transverse suture either complete or interrupted in the middle. Scutellum large, rounded posteriorly, with two long setae. Postnotal calli rounded, with a vertical row of setae. *Wings*: Two cross veins, anal cross vein absent. Cross vein i–m markedly closer to base of wing than r–m, so that the second basal cell is much shorter than the first. Membrane covered with microtrichia in its greater part. *Legs*: Claws double. Pulvilli normal. Empodium feathered. Basitarsus of hind legs longer than that of the other legs.

Abdomen: Membranous in its greater part. Middle of dorsum finely striated. Tergal plates 3 and 5 small, or one or the other absent. Tergal plate 4 always absent. Tergal plate 6 large, rarely divided. Tergite 7 with small lateral sclerites or absent in the female, always absent in the male. Sternite 1 absent. Ventral side completely membranous.

Genitalia: Aedeagus conical, strongly curved at the base. Postgonites triangular, curved. Praegonites broadly rounded, with numerous setae. Anal sclerite of

52

varying form, sometimes Y- or U-shaped. Genital sclerite of female with rounded dorsal tubercles and plates that cover the genital opening.

Icosta is the largest genus of Hippoboscidae, with about fifty-four species that are distributed throughout the world, mainly in the tropics. Some Ethiopian species may reach the Mediterranean occasionally. Three species in Israel.

Key to the Species of Icosta in Israel

1. Scutellum with two long setae near the middle of the posterior margin. Ventral side of head with numerous long hairs, particularly in the male. Palps long, triangular in profile. On Falconiformes. **I. meda** Maa
 – Scutellum with a long seta at each side. Palps shorter   2

2. Larger species. Wing length 6 mm. Colour brown. Frontal horns forming a narrow V. Palps very short, as long as the antennae. Second basal cell half as long as the first. Microtrichia cover the wing membrane beyond the anal vein. On Ciconiiformes. **I. ardeae** Macquart
 – Very small, pale yellowish gray species. Wing length 3.5–4.0 mm. Frontal horns widely diverging. Palps longer, as long as the mediovertex. Second basal cell a third as long as the first. Microtrichia cover the wing membrane only to half of cell m_4. On Passeriformes. **I. minor** Bigot

Icosta ardeae (Macquart, 1835)
Figures 103–107

Olfersia ardeae Macquart, 1835. *Histoire naturelle des insectes,* II, p. 640.
Olfersia botauri Rondani, 1879. *Boll. Soc. Ent. Ital.,* 11:22.
Ornithoponus massonnati Falcoz, 1926. O'Mahony (1940) *J. Soc. Br. Ent.,* 2:75 (misdetermination)
nec *Olfersia albipennis* Say, 1823. *J. Acad. Nat. Sci. Philad.,* 3:101.
Lynchia albipennis Say, 1823. Theodor & Oldroyd (1964) in Lindner, 65, Hippoboscidae, p. 47.

The species was for a long time considered as identical with the American species L. *albipennis* Say (Bequaert, 1955), which was thought to be cosmopolitan. Maa (1964) re-established the name *ardeae* for the Old World form, as he found constant differences between it and the American form.

Head + thorax 3.25 mm. Wing length 6.0–6.5 mm. Colour dark brown.

Head: Broadly oval, with nearly straight posterior margin. Eyes moderately large. Vertex nearly twice as wide posteriorly as an eye. Parafrontalia broad, with indistinct inner margin. A long orbital seta anteriorly, one in the middle and several rows of fine yellowish hairs. Mediovertex narrower anteriorly. Postvertex trapezoidal, with indistinct anterior margin. One pair of vertical setae, which stands distinctly behind the posterior margin of the eyes. Ocelli absent. Lunula distinctly separated from anterior frons. Anterior frons V-shaped,

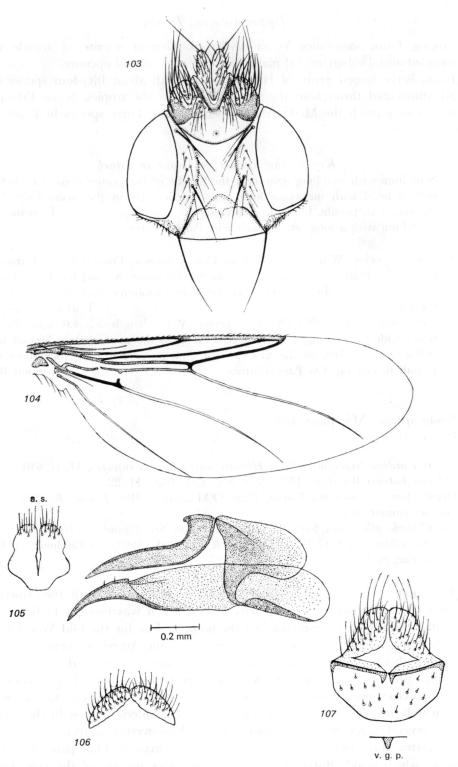

Figs. 103–107: *Icosta ardeae* (Macquart). 103. head, dorsal; 104. wing; 105. male genitalia, with anal sclerite (a.s.); 106. female, anal frame, dorsal; 107. genital sclerite of female, with ventral genital plate (v.g.p.)

narrow. Basal segment of antennae light, second segment dark. Palps short, as long as the antennae, with rounded end in profile.

Thorax: Humeral calli conical, rounded, with spines at the outer surface and a long seta dorsally. Anterior spiracle directed laterally. Median longitudinal suture of mesonotum distinct. Transverse suture complete or interrupted in the middle. Only fine yellow hairs on the anterior part of the mesonotum, before the scutellum and on it. Mesonotum wrinkled. Scutellum broadly rounded posteriorly, with straight anterior margin and with a longitudinal groove in the middle. A long seta at each side. Postnotal calli rounded, low, with a vertical row of four to six black setae. Prosternum bilobed, with two long setae. *Wings*: Two cross veins. R_2+_3 ends near R_4+_5 in the costa. Second basal cell half as long as the first. Microtrichia cover the membrane beyond the anal vein. *Legs*: Basitarsus of hind legs as long as tarsi 2–4 together. Pulvilli normal.

Male Abdomen: Tergite 1+2 with straight posterior margin, with dense setae laterally. Tergal plates 3–5 absent. Tergal plate 6 wide, sometimes concave or incised at the posterior margin, about half as wide as the abdomen, with long setae laterally at the posterior margin. Middle of dorsum transversely striated. Lateral parts of tergite 3 sclerotized in older specimens. Sternite 1 absent. *Genitalia*: Aedeagus 0.6–0.7 mm long, strongly curved at the base, conical, pointed. Postgonites triangular, curved. Praegonites broadly rounded, with numerous setae. Anal sclerite U-shaped.

Female Abdomen: As in the male, but tergal plate 3 present, small, elliptical, sometimes covered by tergite 1+2. A row of long, strong setae on sclerotized tubercles before the genital opening. Genital sclerite with rounded dorsal process, which is divided and covered with setae (Fig. 107).

Hosts: Ciconiiformes, mainly small species of Ardeidae.

Distribution: Tropics and subtropics from 30° lat. S. to 54° lat. N.; Mediterranean; rarely C. Europe and British Isles; N. Asia to Japan.

Israel: Galilee, Yezrc'el Valley (15), northern Jordan Valley, on *Ardea cinerea* L., *A. purpurea* L., *Ardeola ralloides* (Scop.), *Botaurus stellaris* L.; about twenty specimens examined; February to June.

Icosta meda (Maa, 1963)
Figures 108–113

Lynchia meda Maa, 1963. *Pacif. Insects,* Monograph VI, p. 161.
Lynchia barbata Theodor & Oldroyd (1964) in Lindner 65, Hippoboscidae, p. 48.
Lynchia interrupta Maa, 1964. *J. Med. Ent.,* 1 : 100.

Head+thorax 3.5 mm. Wing length 6.5–7.0 mm. Colour dark brown.
Head: Broadly oval. Posterior margin rounded, vertex nearly twice as wide as an eye. Parafrontalia parallel-sided, dark brown, nearly black laterally, with a long seta anteriorly, a long seta behind the middle and fine yellow hairs.

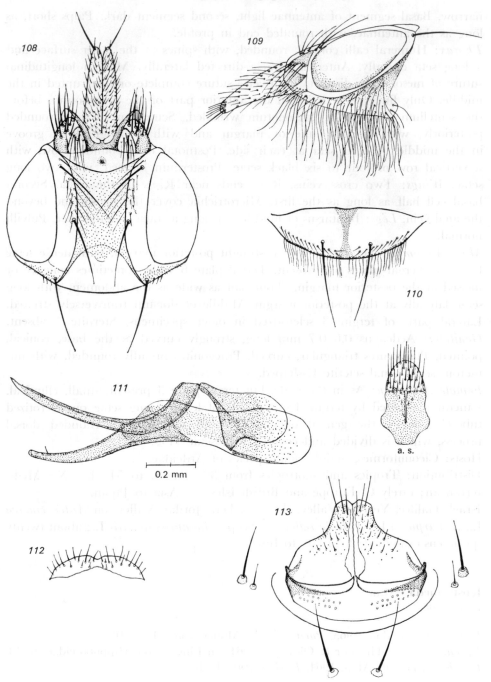

Figs. 108–113: *Icosta meda* (Maa). 108. head, dorsal; 109. head, lateral;
110. scutellum; 111. male genitalia, with anal sclerite (a.s.); 112. female,
anal frame, dorsal; 113. female, genital sclerite and setae near
genital opening

Postvertex trapezoidal, with an incision anteriorly, longer than in *I. ardeae*. One pair of long, yellow vertical setae. Anterior frons with broad base and diverging horns. Ventral side of head with numerous long, yellow hairs, particularly in the male. Two to three long jugular setae anteriorly in the female. Palps as long as the mediovertex, triangular in profile.

Thorax: Humeral calli pale, blunt, rounded-conical, with short spines anteriorly and laterally and one or two long setae dorsally. Anterior spiracle broadly elliptical, on the dorsal surface of the mesonotum at the base of the humeral callus. Adjacent part of mesonotum also pale. Mesonotum anteriorly with an oblique row of long, yellow setae. Setae on humeral calli and notopleuron black, all other setae on mesonotum yellow. Longitudinal suture of mesonotum distinct, transverse suture interrupted in the middle. Scutellum rounded posteriorly, with a broad median groove and two long setae near the middle of the posterior margin. Postnotal calli low, rounded, with a vertical row of four to six black setae. Prosternum not bilobed, with several yellow setae. Praesternum large, triangular. *Wings:* Venation as in *I. ardeae*, but microtrichia cover only half of cell m₄.

Male Abdomen: Tergal plates 3–5 absent. Dorsum striated in the middle. Tergal plate 6 broad, with concave posterior margin or divided ('*interrupta*') and long black setae posteriorly at the sides. *Genitalia:* Aedeagus 0.8 mm. long, very slender, tapering, curved at the base. Postgonites slender, slightly curved. Praegonites oblong, narrower than in *I. ardeae*, with numerous setae.

Female Abdomen: As in the male, but tergal plate 3 present, small, elliptical or rectangular. Tergal plates 4 and 5 absent. Tergal plate 6 wide, strip-like, with four to six long, black setae at the sides of the posterior margin and short spines in the middle. Sclerotized lateral parts of tergite 3 extend to the pleurae. Area before the genital opening bare, except for two long setae and a long and a short seta lateral to the genital opening. Genital sclerite with narrow dorsal process, divided, with short setae (Fig. 113).

Hosts: *Aegypius monachus* L., *Gyps fulvus* (Habl.).

Distribution: Iraq; Africa.

Israel: Be'er Sheva' area (15), twenty specimens; *Aegypius monachus* and *Gyps fulvus*; June to August.

Icosta minor (Bigot, 1858)
Figures 114–115

Olfersia minor Bigot, 1858. Thomson, *Archs Ent.*, 2 : 376.
Olfersia falcinelli Rondani, 1879. Bequaert (1955) p. 366.
Olfersia minor Bigot. Speiser (1902) *Z. syst. Hymenopt. Dipterol.*, 2 : 156.
Lynchia minor Bigot. Theodor & Oldroyd (1964) in Lindner, 65, Hippoboscidae, p. 51.

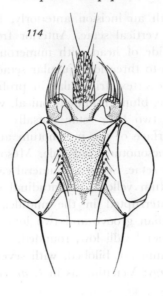

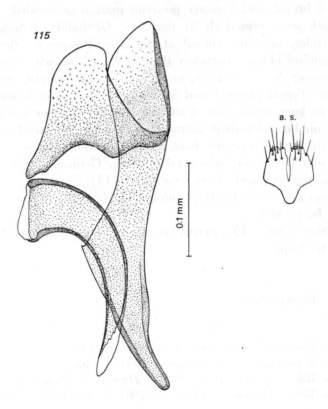

Figs. 114–115; *Icosta minor* (Bigot). 114 head, dorsal, 115. male genitalia, with anal sclerite (a.s.)

Head+thorax 2 mm. Wing length 3.5–4.0 mm. Colour pale yellowish gray.

Head: Broadly rounded. Vertex nearly twice as wide as an eye. Parafrontalia broad, parallel-sided, with a long yellow seta anteriorly, one seta in the middle and with fine yellow hairs. Mediovertex twice as wide as a parafrontal. Postvertex trapezoidal, wider than long. Horns of anterior frons widely diverging. Palps as long as the mediovertex.

Thorax: Humeral calli conical, with two to three long setae. Transverse suture of mesonotum narrowly interrupted in the middle; longitudinal suture distinct. Anterior spiracles small, situated on the dorsal surface of the mesonotum. Praescutellar and notopleural setae thin, yellow. Setae on humeral calli, mesopleura and postalar setae black. A diagonal row of five to six yellow setae behind the humeral suture. Scutellum with rounded posterior margin and with a long, black seta at each side. Postnotal calli low, rounded, with a few black setae. *Wings*: Costa markedly thickened apically. Second basal cell very short, about one-third as long as the first. Membrane covered with microtrichia to half of cell m₄.

Male Abdomen: Tergite 1 + 2 very short, with straight posterior margin. Tergal plates 3–5 absent. Tergal plate 6 wide, with two to three setae at the sides of the posterior margin. Dorsum striated in the middle. *Genitalia*: Aedeagus short and thick, conical, curved at the base, membranous at the end. Postgonites triangular. Praegonites rounded, with setae at the margin. Anal sclerite divided, with short setae at the ends.

Female Abdomen: As in the male, but tergal plate 6 with concave posterior margin and long setae at the sides of the posterior margin. Genital sclerite with rounded dorsal process and broad plates which cover the genital opening.

Hosts: Passeriformes, mainly Laniidae.

Distribution: Africa; Mediterranean; N.W. Russia to Middle Asia.

Israel: Mishmar Ha'Emeq (5), one female from *Lanius senator* L.; 28 May 1956. Hazeva, 'Arava (14), one female from *Turdoides squamiceps* (Cretzschm.); 22 January 1972.

Diptera Pupipara

Genus PSEUDOLYNCHIA Bequaert, 1926
Psyche, 32 : 271.

Lyncia Speiser, 1902. *Z. syst. Hymenopt. Dipterol.*, 2 : 155.

Type Species: *Olfersia canariensis* Macquart, 1840, as *Olfersia maura* Bigot, 1885.

Head: Broadly rounded, with nearly straight posterior margin. Eyes moderately large, but posterior orbits distinct, triangular. Vertex one and a half times to twice as wide as an eye. Parafrontalia broad, with a long seta anteriorly, one seta in the middle and with numerous fine, yellow hairs. Postvertex wider than long, without ocelli. One pair of vertical setae. Anterior frons with broad base and widely diverging horns. Antennae with short dorsal process. Arista spatula-shaped. Palps as long as mediovertex or shorter.

Thorax: Markedly dorso-ventrally depressed, so that the mesopleura are broadly visible dorsally. Humeral calli long, conical. Mesonotum with diagonal rows of long setae anteriorly. Notopleuron reduced, but with a long seta on a tubercle. Scutellum rectangular or slightly rounded, with finger-shaped processes at the sides of the posterior margin. Postnotal calli rounded, with a vertical row of setae. Prosternum with two triangular processes. *Wings*: Only cross vein r–m present, so that there is no second basal cell and no anal cell. Wings covered with microtrichia in their greater part. *Legs*: Claws double. Pulvilli normal. Empodium feathered.

Abdomen: Tergite 1+2 with straight posterior margin. Tergal plate 3 small. Tergal plate 4 absent. Tergal plate 5 reduced or absent in the male, absent in the female. Tergal plate 6 broad, with concave posterior margin. Tergal plate 7 absent.

Genitalia: Aedeagus curved near the base, widened in the form of an axe at the end. Postgonites very slender, rod-shaped. Anal sclerite Y-shaped, with setae at the ends. Genital sclerite of female with rounded dorsal process which is not divided. Ventral genital plate very small.

Five species distributed in the tropics and subtropics. One species is cosmopolitan, one American, two occur in New Guinea, and an African species (*P. garzettae*) occurs rarely in the Mediterranean. One species in Israel, but *P. garzettae* may occur occasionally.

Key to the Palaearctic Species of Pseudolynchia

1. Vertex one and a half times as wide as an eye. Palps as long as the mediovertex. Frontal horns widely diverging, with a rounded invagination between them. Base of anterior frons as broad as the distance from its posterior angle to the eye. A double, diagonal row of twenty to thirty long, yellow setae anteriorly on the mesonotum. Scutellum rectangular. Posterior process of basisternum 3 above coxa 3 long, pointed. A group of peg-like setae on the basitarsus of the mid-leg of the male. Unspecific, but mainly on Columbiformes.

   P. canariensis Macquart

– Vertex twice as wide as an eye. Palps distinctly shorter than mediovertex, as long as the frons. Frontal horns form a V, its base narrower than its distance to the eye. A single diagonal row of twelve to eighteen long setae which may be dark, on the anterior part of the mesonotum. Scutellum with slightly rounded posterior margin and oblique sides. Posterior process of basisternum 3 short, broadly triangular. Setae on basitarsus of mid-leg of male not peg-like, but spine-shaped. Only on Strigiformes and Caprimulgiformes.

P. garzettae Rondani

Both species are difficult to distinguish, as the characters vary markedly, and *P. garzettae* has, therefore, probably been frequently overlooked in the Mediterranean.

Pseudolynchia canariensis (Macquart, 1840)
Figures 1, 6, 8–9, 31, 116–124, 129

Olfersia canariensis Macquart, 1840. Webb & Berthelot, *Histoire naturelle des Iles Canaries (Entomologie)*, II, p. 119.
Olfersia testacea Macquart, 1843. (1842) *Mém. Soc. Sci. Agric. Lille*, p. 434.
Olfersia rufipes Macquart, 1847. *Ibid.*, p. 229.
Olfersia falcinelli Rondani, 1879. *Boll. Soc. Ent. Ital.*, 11 : 23.
Olfersia maura Bigot, 1885. *Annls Soc. Ent. Fr.*, 4 : 237.
Olfersia capensis Bigot, 1885. *Ibid.*, p. 240.
Pseudolynchia canariensis Macquart. Theodor & Oldroyd (1964) in Lindner, 65, Hippoboscidae, p. 53.

Head + thorax 3.5 mm. Wing length 5–7 mm. Colour grayish brown.
Head: Eyes moderately large. Posterior orbits distinct. Vertex one and a half times as wide as an eye. Parafrontalia broad. Anterior frons with wide base, as wide as the distance to the eye. Palps as long as the mediovertex.
Thorax: Humeral calli whitish, with two long setae. A double diagonal row of twenty to thirty long, yellow setae on the anterior part of the mesonotum. Transverse suture of mesonotum interrupted in the middle. Scutellum rectangular, with a long, black seta at each side. *Wings*: Microtrichia cover the membrane to the middle of cell m$_4$. Posterior process of basisternum 3 above coxa 3 long and pointed. *Legs*: An irregular group of peg-like setae on the basitarsus of the mid-legs of the male.
Abdomen: As described for the genus. Tergal plate 3 of the male larger than in the female. Tergal plate 5 of the male may be absent or reduced to small lateral sclerites.
Genitalia: Aedeagus 0.85 mm long. Postgonites very slender, rod-like. Praegonites broadly rounded, with numerous setae. Genital sclerite of female with rounded, undivided dorsal process and a broad plate inside the genital opening. Ventral genital plate very small, triangular.
Hosts: Columbiformes, but also on many other birds in the Old World; only on the domestic pigeon in America.
Distribution: Tropics and subtropics of the whole world, rare in C. Europe.

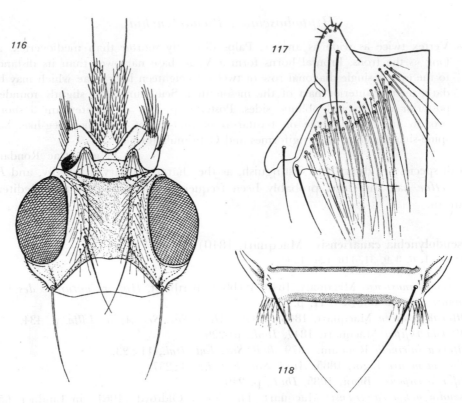

Figs. 116–118: *Pseudolynchia canariensis* (Macquart). 116. head, dorsal
(after Jobling, 1926); 117. setae on anterior part of mesonotum;
118. scutellum

Israel: Galilee to Be'er Sheva' (15); common on domestic pigeon; also recorded
from *Columba livia* L., *Streptopelia turtur* L., *S. decaocto* (Friv.), *Alectoris
graeca* (Meisner), *Milvus migrans* (Bodd.), *Upupa epops* L., *Nycticorax nycticorax* L., *Cuculus canorus* L., *Corvus cornix* L.; April to October.

Pseudolynchia garzettae (Rondani, 1879)
Figures 125–128, 130

Olfersia garzettae Rondani, 1879. *Boll. Soc. Ent. Ital.*, 11:23.
Pseudolynchia fradeorum Tendeiro, 1951. *Centro Estudios Guinea Portuguesa*,
15 : 127.
Pseudolynchia rufipes Macquart, 1847. Bequaert, 1935; 1953–1957; Theodor &
Oldroyd (1964) in Lindner, 65, Hippoboscidae, p. 54.

The species resembles *P. canariensis* closely and differs in the characters given
in the key. It also differs in its host relationships, as it is mainly a parasite of
Strigiformes and Caprimulgiformes. It is mainly distributed in the Oriental region and in Africa, but also occurs in the Mediterranean, where it apparently
breeds sometimes, as it was found in Cyprus on *Athene noctua* Scop., a resident
bird.

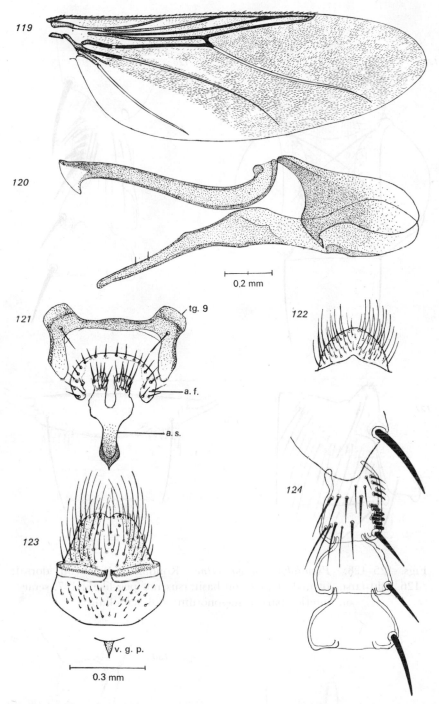

Figs. 119–124: *Pseudolynchia canariensis* (Macquart). 119. wing; 120. male genitalia; 121. male, tergite 9, anal frame and anal sclerite; 122. female, anal frame, dorsal; 123. female, genital sclerite, and ventral genital plate (v.g.p.); 124. mid-tarsus of male, with peg-like spines on the basitarsus
a.f. – anal frame; a.s. – anal sclerite; tg.9. – tergite 9

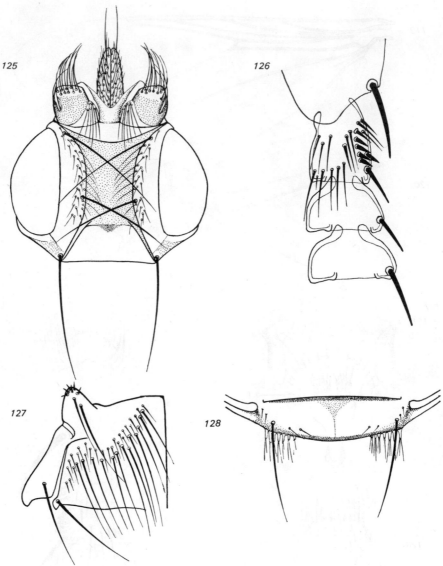

Figs. 125–128: *Pseudolynchia garzettae* (Rondani). 125. head, dorsal;
126. mid-tarsus of male (spines on basitarsus not peg-like); 127. setae
on anterior part of mesonotum; 128. scutellum

Figs. 129–130: Posterior process of basisternum 3 above coxa 3.
129. *P. canariensis* (Macquart); 130. *P. garzettae*

Genus O L F E R S I A Wiedemann, 1830
Aussereuropäische zweiflüglige Insekten, II, p. 605.

Feronia Leach, 1817. *On the Genera and Species of Eproboscideous Insects,*
pp. 3–11 (preoccupied).
Pseudolfersia Coquillett, 1899. *Can. Ent.,* 31 : 336.

Type Species: *Feronia spinifera* Leach, 1817.
Head: Flattened, broadly oval. Eyes large. Vertex parallel-sided. Parafrontalia
narrow. Posterior orbits present. Posterior margin of head with three broadly
rounded bulges. Postvertex very long, nearly reaching the frontal suture, so
that the mediovertex is markedly reduced. Ocelli absent. Lunula large, with
a large pit in the middle, fused without suture with the anterior frons. Horns of
anterior frons broad and long, striated, contiguous in the middle. Anterior point
of horns directed laterally. Dorsal process of antennae rounded. Basal segment of
antennae distinctly divided from the lunula. Arista spatula-shaped.
Thorax: Markedly flattened, chaetotaxy markedly reduced. Prothorax not
visible dorsally. Humeral calli broadly conical, blunt. Anterior margin of mesonotum with two rounded protuberances. Median longitudinal suture of mesonotum distinct to the scutellum; transverse suture complete or narrowly interrupted in the middle. Notopleuron fused with mesonotum, with one thin seta.
One postalar seta, two praescutellar setae. Scutellum broad, rounded-rectangular, without setae. Postnotal calli large, with a curved posterior process
and fine hairs, but without setae. Prosternum fused with mesosternum, undivided, concave anteriorly. Sternal plate without setae, except for two thin setae
on the prosternum. *Wings*: Two cross veins, anal cross vein absent. Wings covered with microtrichia in their greater part. R_4+_5 and basal part of M_1+_2 with
short setae in some species. *Legs*: Hind tibiae with a row of about six spines at
the apex on the ventral side. Basitarsus of hind leg longer than tarsi 2–4
together.
Abdomen: Membranous in its greater part, transversely striated in the middle
of the dorsum. Tergal plate 3 very narrow and wide, sometimes divided. Tergal
plates 4 and 5 absent. Tergal plate 6 broad, with setae at the posterior margin.
Tergal plate 7 absent. Sternite 1 small, transversely rectangular, with fine hairs.
Anal frame of female with processes which may be fused.
Genitalia: Aedeagus of male broader at the base, more or less parallel-sided in
the apical part. Postgonites long, triangular, slightly curved. Praegonites very
large, finger-shaped in the apical part.
Puparium with T-shaped hairs.
Seven species, some species on oceanic birds, mainly in the tropics. Some species
are narrowly host-specific. One species Palaearctic, another accidental in the
Palaearctic region. One species in Sinai and the north of the Red Sea. The species occur on Pelecaniformes, Procellariiformes, Falconiformes, Galliformes.

65

Olfersia fumipennis (Sahlberg, 1886)
Figures 131–138

Lynchia fumipennis Sahlberg, 1886. *Meddn Soc. Fauna Flora Fenn.*, 13 : 150.
Pseudolfersia maculata Coquillett, 1899. *Can. Ent.*, 31 : 336.
Pseudolfersia mycetifera Speiser, 1905. *Z. syst. Hymenopt. Dipterol.*, 5 : 539.
Olfersia fumipennis (Sahlberg). Bequaert (1957) *Entomologica Am.*, 36 : 444.
Olfersia fumipennis (Sahlberg). Theodor & Oldroyd (1964) in Lindner, 65, Hippoboscidae, p. 55.

Head+thorax 4.0–4.5 mm. Wing length 8–9 mm. Colour brown to black.
Head: Three shallow, rounded bulges at the posterior margin, the lateral bulges with short spines. Parafrontalia narrow, slightly wider posteriorly, with a thin seta anteriorly and a few thin, shorter setae to the middle. One pair of vertical setae. Postvertex large, rounded-triangular anteriorly, shining black, reaching to near the lunula. Horns of anterior frons broad, longitudinally striated, forming an obtuse angle anteriorly. Palps short.
Thorax: Humeral calli large, with rounded apex, with spines at the apex and with two setae dorsally. Transverse suture complete, angular posteriorly, tomented. Two tomented areas before the scutellar suture around the praescutellar setae. Scutellum broad, rounded-rectangular, but with more strongly convex anterior margin, without setae. Postnotal calli with a curved posterior process. *Wings*: Section of costa between Sc and R_1 longer than section between R_1 and R_{2+3}. Sections 2 : 3 : 4 of costa = 2 : 1.5 : 1. First basal cell shorter than part of R_{4+5} distal to r–m. Second basal cell broader at the base. R_{4+5} with some light, short setae its whole length. The microtrichia cover the wing to the middle of cell m_4, reaching the anal vein only near the base.
Male Abdomen: Tergite 1+2 with slightly concave posterior margin, tomented, except in a stripe in the middle of the posterior margin. Tergal plate 3 very wide and narrow, sometimes divided. Tergal plate 6 large, trapezoidal, two and a half times wider than long, about half as wide as the abdomen. Posterior margin concave, with a row of long setae. *Genitalia*: Aedeagus 0.9 mm long, thicker at the base, nearly parallel-sided in the middle, wider and obliquely truncate at the apex and with a few minute denticles at the dorsal margin. Postgonites long, triangular, slightly curved. Praegonites very large, finger-shaped in the apical part, with numerous setae. Anal sclerite strip-shaped, with a few short setae at the apex.
Female Abdomen: Tergal plate wider than in the male, more than half as wide as the abdomen, often divided. Pleurae of tergite 3 with a moderately large, rounded sclerotization. Tergal plate 6 narrower than in the male, four to five times wider than long, its posterior margin straight or slightly convex. Ventral side of abdomen covered with short setae in the greater anterior part. Two or three rows of long setae before the genital opening. A bare area between the anterior setose area and the setae near the genital opening. Two small sclerites with two or three long setae lateral to the opening. Anal frame with fused

66

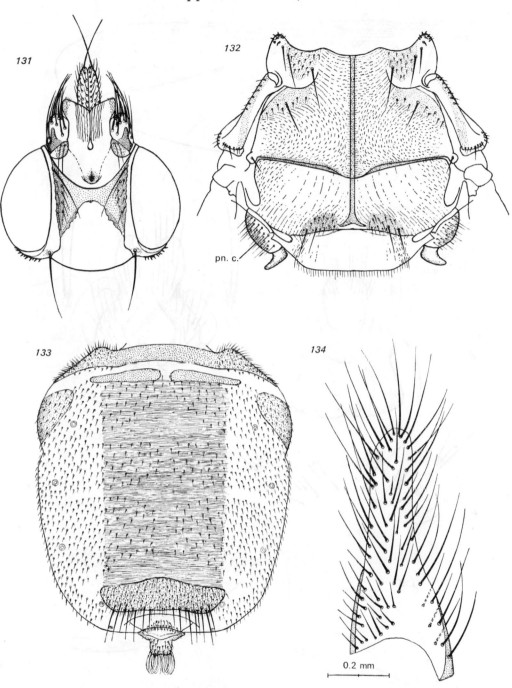

Figs. 131–134. *Olfersia fumipennis* (Sahlberg). 131. head, dorsal;
132. thorax, dorsal; 133. female abdomen, dorsal;
134. male, praegonite
pn.c. – postnotal callus

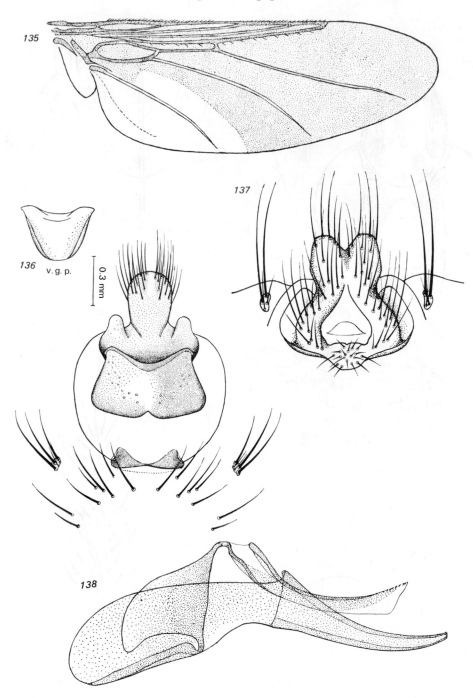

Figs. 135–138: *Olfersia fumipennis* (Sahlberg). 135. wing; 136. female, genital sclerite and setae near genital opening; ventral genital plate (v.g.p.); 137. female, anal frame, dorsal; 138. male genitalia

processes, sometimes with an indentation at the apex. Genital sclerite as in Fig. 136.

Hosts: *Pandion haliaëtus* L., rarely *Haliaëtus leucocephalus* L. or other birds. Distribution: Nearly cosmopolitan with *Pandion haliaëtus* to 63° lat. N. in Finland. Recorded from the north of the Red Sea (Quseir, Sinafir Island in the Straits of Tiran).

Sinai: One female, Sharm e-Sheikh (23) from *Pandion haliaëtus*; 2 February 1969.

Pandion haliaëtus occurs in Israel; *Olfersia fumipennis* therefore probably also occurs there.

Subfamily LIPOPTENINAE Speiser, 1908

MELOPHAGINAE Bezzi, 1916, *Revue Sci. Nat.,* 7 : 177; Bequaert (1942) p. 36; Theodor & Oldroyd (1964) in Lindner, 65, Hippoboscidae, p. 56.

DIAGNOSIS

Head triangular ventrally, trapezoidal dorsally, deeply fitted into the thorax, so that lateral movement is markedly restricted. Anterior margin of frons smoothly rounded. Transverse suture of mesonotum always broadly interrupted in the middle, the two lateral halves directed posteriorly and inwards to near the scutellar suture, or suture absent (*Melophagus*). Wings and halteres absent in *Melophagus,* functional at hatching, with only three longitudinal veins, breaking off near the base when the host is reached in the other two genera. Claws simple. Parasites only of Artiodactyla. Three genera, of which two are Palaearctic, both represented in Israel.

Genus LIPOPTENA Nitzsch, 1818
Germar's Mag. Ent., 3 : 310.

Haemobora Curtis, 1824, *British Entomology,* I, Pl. XIV.
Ornithobia Meigen, 1830. *Systematische Beschreibung der Europäischen zweiflügligen Insekten,* VI, p. 229.
Alcephagus Gimmerthal, 1845. *Stettin Ent. Ztg,* 6 : 152.

Type Species: *Pediculus cervi* Linnaeus, 1758.
Head: Nearly trapezoidal dorsally, with broadly rounded anterior margin, triangular ventrally. Eyes small, more or less elliptical or strip-like, extending to the anterior side of the head. Parafrontalia broad. Mediovertex large. Postvertex rounded. Ocelli present or absent. Antennal pits closed, situated at the sides of the frons. Antennae small, rounded, without dorsal process. Arista branched or spatula-shaped. Palps shorter than frons.

Thorax: Humeral calli rounded, not projecting. Anterior spiracle large, broadly oval, at the lateral margin of the mesonotum. Median longitudinal suture of mesonotum short or reaching to the scutellar suture. Transverse suture consisting of lateral halves which run obliquely inwards to near the scutellar suture. Notopleuron small, indistinctly divided from the mesonotum. Scutellum small, rounded posteriorly, with setae at the posterior margin. Usually a row of acrostichal setae, a varying number of laterocentral setae, a double row of mesopleural setae, a row of postalar setae, a row of praescutellar setae and a row of scutellar setae. Posterior spiracle situated at the base of the postnotal callus, directed posteriorly. Postnotal calli rounded, with a few setae. Prosternum large, divided into triangular, widely separated processes, between which the posterior ventral process of the head is fitted. Basisternum 2 large, covered with setae, basisternum 3 smaller, with setae only in the posterior half. Furcasternum 3 large, with a median incision posteriorly, without setae. *Wings*: Normally developed at hatching, functional, with only three longitudinal veins, R_1, R_{4+5} and M_{3+4}. The wings break off near the base when the host is reached. Halteres normal. *Legs*: Short and thick, particularly the fore femora. Claws simple, asymmetrical. One or both pulvilli rudimentary. Empodium with very short hairs.

Abdomen: Membranous in its greater part. Tergal plates more or less reduced. Tergite 1+2 consists of two median, strongly sclerotized parts and less sclerotized lateral parts which reach far posteriorly in some species. Tergal plates 3–5 large or small, tergal plate 6 larger, sometimes divided. Tergal plate 7 of female divided into lateral sclerites, absent in the male. Tergal plates 3 and 4, or 3–5, absent in some species. Seven pairs of abdominal spiracles, 3–5 in the membrane, 6 and 7 in the tergal plates or near them in the membrane. Spiracle 5 may be markedly displaced posteriorly in extended females of some species. Sternite 1 with lateral posterior processes covered with spines. Ventral side otherwise membranous.

Genitalia: Aedeagus tubular, with a ventral endophallus or divided into a dorsal and a ventral part. Postgonites triangular. Praegonites rounded, with numerous setae. Genital sclerite of female with a dorsal plate with numerous setae, and with narrow plates which cover the genital opening and a second plate inside the opening. One to three ventral genital plates.

Twenty-nine species, of which nine are Palaearctic and three American. Hosts only Artiodactyla. Two species in Israel.

Key to the Species of Lipoptena in Israel

1. Two to three long orbital setae. Ten to fifteen laterocentrals and a bare area between them and the acrostichals. One pulvillus rudimentary. Ten to twelve setae in two rows before the genital opening of the female. Aedeagus very slender, 1.6 mm long, three times as long as the postgonites.

   L. capreoli Rondani

– Four to eight orbital setae. Fifteen to twenty-five laterocentrals, so that the mesonotum is completely covered with setae, without a bare area between laterocentrals and acrostichals. Both pulvilli rudimentary. Thirty to forty setae in four to five rows before the genital opening of the female. Aedeagus shorter, 1.3 mm long, twice as long as the postgonites.   **L. chalcomelaena** Speiser

Lipoptena capreoli Rondani, 1879
Figures 13, 23–24, 32, 37, 139–142, 144, 146–148

Lipoptena caprina Austen, 1921. *Bull. Ent. Res.*, 12 : 122.
Lipoptena capreoli Rondani. Theodor & Oldroyd (1964) in Lindner, 65, Hippoboscidae, p. 59.

Head+thorax 1.8–2.3 mm. Wing length 3.0–3.2 mm.
Head: Eyes narrow, not reaching to the sides of the head. Parafrontalia posteriorly as broad as an eye. Two or three long orbital setae and a few shorter setae. Postvertex large, nearly twice as wide at the base as long. One pair of long vertical setae. The ocelli form an equilateral triangle. Palps short, as long as the antennal pits.
Thorax: Eight to ten acrostichals in a curved row and ten to fifteen laterocentrals, leaving a bare space between them and the acrostichals. Both halves of the transverse suture of the mesonotum reach to near the scutellar suture; one praescutellar seta, three or four postalar setae at each side, widely separated from the praescutellar setae. Scutellum with six setae at the posterior margin. *Wings*: R_1 nearly parallel to R_{4+5}, ending far before the apex of the wing. Costa thin to R_1, markedly thicker between R_1 and R_{4+5}. *Legs*: Claws asymmetrical. Only one pulvillus at the shorter claw.
Male Abdomen: Tergite 1+2 divided in the middle, the inner parts angular posteriorly. Lateral parts relatively short. Tergal plates 3 and 4 small, tergal plate 5 larger. Tergal plate 6 large, rectangular. Tergal plate 7 absent. Sternite 1 with markedly concave posterior margin, with long spines at the posterior processes and shorter spines on the surface and at the concave part of the posterior margin. *Genitalia*: Aedeagus long and slender, 1.6 mm long. Postgonites triangular, relatively small, slightly curved. Anal sclerite triangular, with numerous setae posteriorly.
Female Abdomen: Inner parts of tergite 1 + 2 shorter than in the male, rounded posteriorly, lateral parts longer. Tergal plates 3–5 smaller than in the male, 4 smaller than 3 and 5. Tergal plate 6 large, rectangular, with two setae laterally at the posterior margin. Tergal plate 7 divided into small lateral sclerites with two setae. Spiracles 6 and 7 in the membrane near the tergal plates. Two rows of setae before the genital opening. Genital sclerite with rounded dorsal plate. The abdomen of extended females forms two posterior processes in which spiracle 5 is situated (Fig. 23).
Hosts: Domestic goats, rarely wild goats. Occasionally camels, cattle, dogs.
Distribution: E. Mediterranean to N.W. India.
Israel: Galilee to Be'er Sheva' (15); common on goats; April to November.

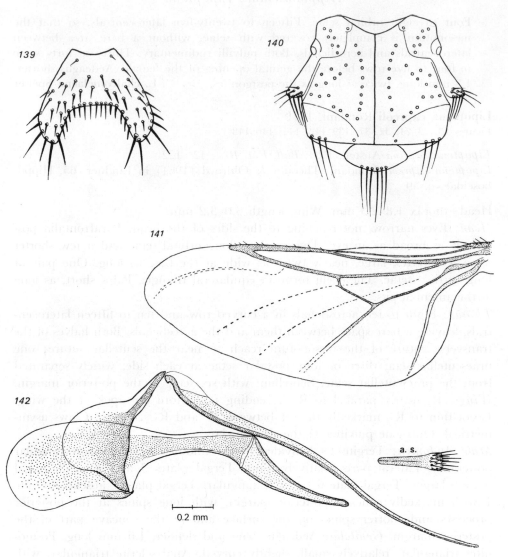

Figs. 139–142: *L. capreoli* Rondani. 139. basal sternite of abdomen;
140. thorax, dorsal; 141. wing; 142. male genitalia, and anal
sclerite (a.s.)

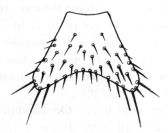

Fig. 143: *Lipoptena chalcomelaena* Speiser, basal sternite of abdomen

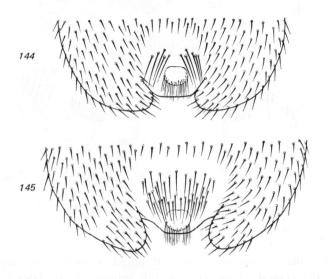

Figs. 144–145: Female abdomen, ventral, posterior part. 144. *Lipoptena capreoli* Rondani; 145. *L. chalcomelaena* Speiser

Lipoptena chalcomelaena Speiser, 1904
Figures 143, 145, 149–151

Lipoptena chalcomelaena Speiser, 1904. *Z. syst. Hymenopt. Dipterol.*, 4 : 178.
Lipoptena ibicis Theobald, 1906. *Rep. Wellcome Trop. Res. Labs*, 2 : 88.
Lipoptena chalcomelaena Speiser. Theodor & Oldroyd (1964) in Lindner, 65, Hippoboscidae, p. 61.

The species is closely related to *L. capreoli* and differs in chaetotactic characters and in the genitalia as given in the key. The uniform cover of the mesonotum with setae, without a bare space between acrostichals and laterocentrals, is the most striking character. Wings as in *L. capreoli*. Setae on dorsum of abdomen shorter and thicker than in *L. capreoli*, setae on posterior margin of tergites longer than in *L. capreoli*. Sternite 1 of abdomen less deeply concave than in *L. capreoli*. Aedeagus 1.3 mm long, thicker at the base than in *L. capreoli;* postgonites longer and broader than in *L. capreoli*. Anal sclerite narrow, parallel-sided, with a few short setae at the end.
Host: Apparently a specific parasite of *Capra ibex nubiana* Cuvier.
Distribution: Sudan.
Israel and Surroundings: Red Sea area, Dead Sea area and adjacent areas of the Negev. 'En Gedi (13), about twenty specimens, on *Capra ibex nubiana*, 8 April 1953; about fifty specimens, same host, 20 February 1970; Nahal Ze'elim (13), five young, winged specimens, 26 February 1970.

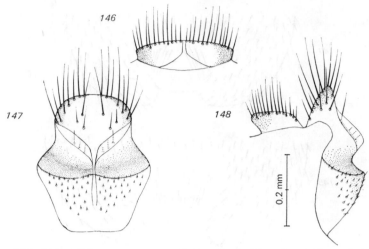

Figs. 146–148: *Lipoptena capreoli* Rondani. Female. 146. anal frame;
147. genital sclerite, ventral; 148. genital sclerite, lateral

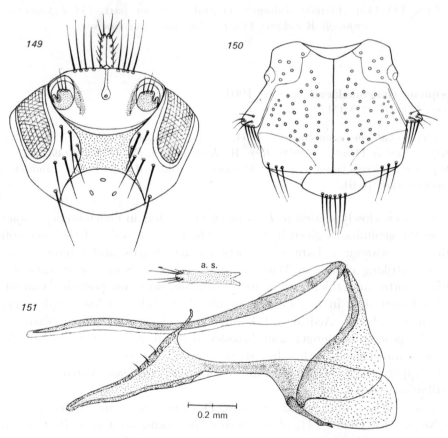

Figs. 149–151: *Lipoptena chalcomelaena* Speiser. 149. head, dorsal;
150. thorax, dorsal; 151. male genitalia, and anal sclerite (a.s.)

Genus MELOPHAGUS Latreille, 1802

Histoire naturelle des crustacées et des insectes, III, p. 466.

Type Species: *Hippobosca ovina* Linnaeus, 1758.

Head: Broadly elliptical dorsally, triangular ventrally, deeply fitted into the thorax. Eyes reduced to a narrow stripe with less than 150 facets. Parafrontalia broadly triangular, reaching nearly to the middle of the mediovertex in some species. Postvertex very large, so that the mediovertex is markedly reduced in size. One or several vertical setae at each side. Frons large, elliptical, without suture between anterior frons and lunula. Antennal pits widely separated. Antennae short. Arista branched. Palps as long as the head or short.

Thorax: Prothorax rounded posteriorly. Humeral calli rounded, not projecting. Median longitudinal suture of mesonotum distinct. Other sutures of mesonotum weakly marked or absent. Scutellum small, elliptical, with several setae at the posterior margin. Postnotal calli absent. Both thoracic spiracles large, the anterior spiracle directed laterally, the posterior spiracle posteriorly. Pleurae partly membranous. Prosternum with two widely separated, triangular processes. Basisternum 2 and 3 with setae. *Wings*: Reduced to small sclerotized stumps with a few setae. Halteres absent. *Legs*: short and thick, tarsi very short. Claws asymmetrical. Pulvilli rudimentary. Empodium feathered.

Abdomen: Nearly completely membranous, uniformly covered with setae. Tergite 1+2 divided. Tergal plates 3–5 absent in both sexes. Tergal plates 6 and 7 present or one or the other absent. Abdominal spiracles large, spiracle 6 and 7 in the membrane near the anus.

Male Genitalia: Aedeagus slightly curved, with large endophallus. Praegonites rounded, with numerous setae.

Puparium: With one or two pits in which or near which the rounded spiracles are situated. Puparia attached to the hairs of the host.

Three species, of which one is cosmopolitan on domestic sheep, one on *Rupricapra* and *Ibex* in Europe and the third on *Gazella gutturosa* Pall. in Mongolia.

Melophagus ovinus (Linnaeus, 1758)
Figures 14, 17, 39, 152–157

Hippobosca ovina Linnaeus, 1758. *Systema Naturae* (10th ed.), p. 607.
Melophaga hirtella Olfers, 1816. *De Vegetativis et Animatis Corporibus...*, I, p. 99.
Melophagus vulgaris M'Murtrie, 1831. *Cuvier, Animal Kingdom,* IV, p. 323.
Melophagus ovinus var. *fera* Speiser, 1908. *Z. syst. Hymenopt. Dipterol.,* 4 : 444.
Melophagus ovinus Linnaeus. Theodor & Oldroyd (1964) in Lindner, 65, Hippoboscidae, p. 65.

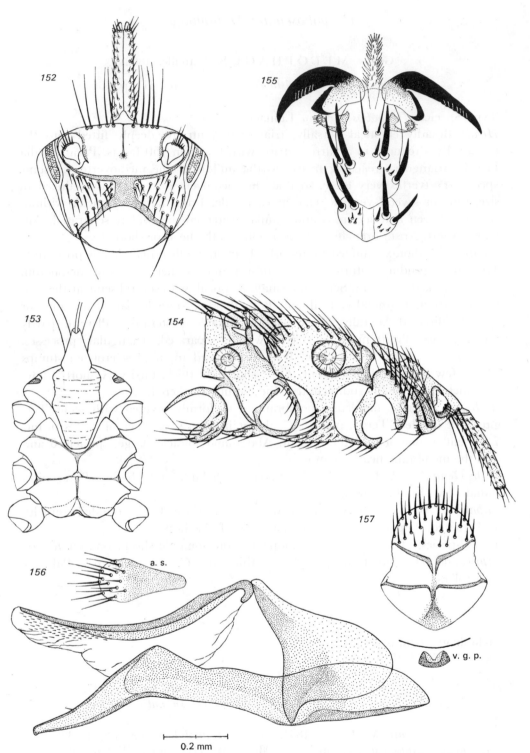

Figs. 152–157: *Melophagus ovinus* (Linnaeus). 152. head, dorsal; 153. head and thorax, ventral; 154. head and thorax, lateral; 155. praetarsus, ventral; 156. male genitalia, with anal sclerite (a.s.); 157. female, genital sclerite and ventral genital plate (v.g.p.)

Head+thorax 2.0–2.5 mm.

Head: Parafrontalia nearly contiguous in the middle, with about twenty setae on the whole surface. One pair of vertical setae. Postvertex triangular, twice as wide at the base as long. Palps nearly as long as the head.

Thorax: Setae on mesonotum irregularly distributed, longer at the posterior margin of the mesopleuron and on the postalar areas. Scutellum rounded, with five to eight setae in the female and eight to ten in the male. Anterior spiracle very large, separated from the mesopleural setae by less than its diameter. Sternopleuron reduced to a narrow stripe.

Male Abdomen: Tergite 1+2 with triangular lateral sclerites. Other tergal plates absent. Dorsum and venter uniformly covered with setae. Sternite 1 triangular, deeply concave posteriorly, covered with short spines and with a few longer setae at the posterior processes. *Genitalia*: Aedeagus curved, 0.75 mm long, with large endophallus. Postgonites triangular. Praegonites small, rounded, with numerous setae. Anal sclerite triangular, broad and rounded posteriorly and with setae in the apical third.

Female Abdomen: Lateral sclerites of tergite 1+2 rounded posteriorly. Tergal plates 3–6 absent. Tergal plate 7 reduced to small sclerites near the anus. Genital sclerite as in Fig. 157.

Hosts: Domestic sheep, accidentally on other hosts, including dogs; bites man occasionally.

Distribution: Europe and Palaearctic Asia to Japan; rare in high altitudes in the tropics (Kenya, Andes); introduced in S. Africa, America, Hawaii, Australia.

Israel and Surroundings: Introduced in the past with sheep from Syria. Numerous specimens from sheep in Ramallah (11) which had been driven in from Syria in April 1926. Occasional records from imported sheep. Recently recorded on local (fat-tailed) sheep in Qiryat Tiv'on (4).

Family STREBLIDAE Kolenati, 1683
Horae Soc. Ent. Ross., 2 : 82.

INTRODUCTION

The Streblidae are pupiparous, blood-sucking, obligatory parasites of bats. Unlike the Nycteribiidae, they contain a surprising variety of forms. Some genera are of the normal habitus of Diptera, with functional wings; others have wings which are reduced to small, non-functional flaps, and one genus is apterous. The American genus *Megistopoda* Macquart resembles Nycteribiidae in habitus, but has reduced wings. Another American genus, *Paradyschiria* Speiser, is apterous and shows modifications of the thorax similar to the Nycteribiidae. The American genus *Nycterophilia* Ferris resembles fleas in the lateral compression of the body; others (*Strebla* Wiedemann, *Metelasmus* Coquillett) resemble Polyctenidae in habitus and development of the ctenidia. The subfamily Ascodipterinae shows the most far reaching adaptation to parasitic life known among Diptera. The female loses wings and legs, burrows into the skin of the bat and is transformed into an unsegmented sac which resembles a trematode more than an insect. The eyes of Streblidae are reduced to several facets, a single facet, or eyes are absent.

The Streblidae are divided into five subfamilies: the Trichobiinae, Streblinae, and Nycterophiliinae are American, and the Nycteriboscinae and Ascodipterinae are distributed in the Old World.

The following description of the morphology refers only to the two Old World subfamilies.

MORPHOLOGY

Head: The head of the Nycteroboscinae is rounded or triangular in profile, funnel-shaped or dorso-ventrally flattened (Figs. 158–159, 161). The head capsule is sclerotized, except for one or two narrow membranous stripes on the vertex and in the gular region. The greater part of the dorsal and lateral surface is occupied by the large laterovertices on which the eyes, if present, are situated. The laterovertices are fused with the postgenae. A small sclerite, the postvertex, is situated posteriorly between the laterovertices, or it is absent. Antennae 2-segmented, scape fused with the head capsule. Second segment with a dorsal slit. Third segment spherical or ovoid, either free or partly invaginated in the second segment (Fig. 160). Arista branched. Palps 1-segmented, flattened, broad, covered with setae. Laterovertices and postvertex covered with

79

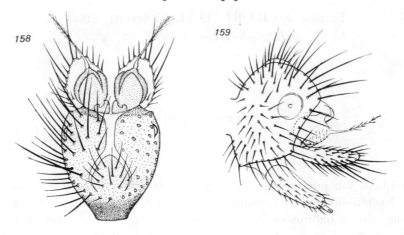

Fig. 158: *Brachytarsina flavipennis* Macquart. Head, dorsal
Fig. 159: *B. alluaudi minor* Theodor. Head, lateral

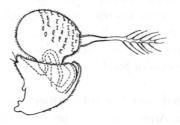

Fig. 160: *B. flavipennis*. Antenna

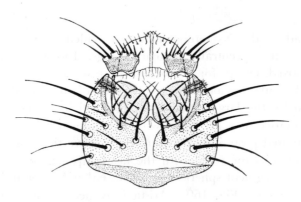

Fig. 161: *Raymondia huberi* Frauenfeld. Head, dorsal

80

setae in varying arrangements. The gular region is surrounded in the genus *Raymondia* by a row of long setae, and by a posterior row of short setae which vary markedly in the different species and give important systematic characters.

The mouth parts resemble those of the blood-sucking Muscidae in principle. The theca of the labium is large, more or less triangular. The labella are short and retracted partly into the theca when not in use. The structure of head and mouth parts has been described in detail by Jobling (1929).

The head of the male of Ascodipterinae resembles that described above in general (Figs. 207, 208, 210). The theca is transversely elliptical and the labella are very short and completely retracted into the theca at rest. They have apparently no praestomal teeth and it is doubtful whether the male sucks blood.

The head of the female (Fig. 197), however, is markedly modified. The head capsule is reduced to a few small sclerites which are connected by wide membranous areas. There is a triangular gena at the sides, and a small occipital sclerite, two laterovertices and the frontal sclerite, which bears the antennae on the dorsal surface. The greater part of the head is formed by the enormous, sclerotized, conical theca of the labium, which bears fourteen rows of long, curved, blade-like teeth on evertible arcs. Palps and eyes are absent.

Thorax: The thorax of the Nycteriboscinae is nearly spherical or dorso-ventrally flattened. Median longitudinal suture of the mesonotum complete, reduced or absent. Postnotal calli with a conical process in some species. Scutellum large, of varying form. Humeral calli present or absent. Sternal plate broad, fore- and mid-legs inserted laterally. It consists mainly of basisterna 2 and 3, which are divided by a median suture. A small prosternum and a small furcasternum 3 are present in some species. The pleural sclerites are fused in their greater part. Pleural suture nearly straight. Legs thick and strong in the Nycteriboscinae, praetarsus triangular, with two strong claws, large pulvilli and a feathered empodium. Mesonotum and sternal plate are covered with setae and microtrichia to a varying extent or microtrichia are absent on some areas. Their distribution gives important systematic characters. The wings have five longitudinal veins and two cross veins in the Nycteriboscinae (Figs. 163, 187). Subcosta absent and anal vein reduced or absent, forming a small anal cell in *Brachytarsina*. Alula well developed, with numerous marginal setae, with one apical seta, or reduced, without setae.

The thorax of the female of Ascodipterinae is laterally compressed, the mesonotum is rounded-rectangular, without longitudinal or transverse sutures. Scutellum small, elliptical. Pleurae with several sclerites which are separated by membranous areas (Fig. 197). The mesopleuron is large and trapezoidal and bears the anterior spiracle. The other sclerites vary in form. The chaetotaxy of these sclerites is systematically important, but it varies markedly in the same species. The sternal plate consists mainly of basisternum 2 with a median suture and an undivided basisternum 3. The prosternum forms a bare transverse stripe of varying width in the female. It is better developed in the male, partly divided and bears setae. The wings *of Ascodipteron* are very delicate. There are only two well developed veins, R_1 and R_{4+5}. R_{2+3} is reduced and the medial and anal veins are slightly indicated by folds in the membrane. Cross veins absent. The whole membrane is covered with microtrichia. Alula absent (Figs. 209–210). Halteres normally developed. The legs are very slender, with slender claws, small pulvilli and a feathered empodium.

Abdomen: The abdomen is nearly unsegmented and membranous in the Nycteriboscinae. Only tergite 1+2 is constantly well sclerotized and bears groups of long setae. Tergal plates 3–5, 3–6 or 3–7 absent. Tergal plates 6 and 7 may be divided into lateral sclerites. A small sternite 1 is present in *Brachytarsina*, absent in *Raymondia*. Sternites 2 and 7 are usually present on the ventral side. The abdomen of the female ends in a sclerotized cone which is mainly formed by segment 10. This apical cone is systematically important, as the form of the sclerites and their chaetotaxy vary widely in different species. The abdomen is covered more or less uniformly with setae laterally and ventrally. The dorsal surface is bare in a wide median space, on which the wings are folded longitudinally between rows of long setae which sometimes stand on sclerotized plates (Fig. 183). These setae protect the wings during movement in the fur of the host. There are six pairs of spiracles; spiracle 1 has been lost.

Male Genitalia (Figs. 172–175, 179–182, 190–195): Those of the Nycteriboscinae are retracted in the abdomen at rest, as in the Hippoboscidae. They consist of a long, slender, tapering aedeagus, more or less rigidly articulated with a long, rod-like, free apodeme which increases markedly in length during the life of the individual. The hypandrium, which has a long apodeme, and the postgonites (parameres) are fused into a rigid sclerite, but the apodeme of the aedeagus is free, and not articulated with the hypandrium as in the Hippoboscidae. The aedeagus is connected with the abdomen by a connective membrane that is short in most species of Brachytarsina, long and tubular but without spines in some other species, and long and covered with spines in *Raymondia*. The hypandrium is connected with the anal frame by a sclerite of varying form, the anal sclerite. This sclerite restricts the extension of the genitalia during copulation. The postgonites vary in form: they are distinctly asymmetrical in *Raymondia*, in which the right postgonite is much larger than the left and bears setae; they are slightly asymmetrical in *Brachytarsina*. The external genitalia are reduced to two small, digitiform processes, the praegonites, which are partly or completely fused in some species.

The abdomen of the female *Ascodipteron* is markedly modified. Only segments 5–7 are indicated by rows of setae and bear large spiracles. Spiracles 1–4 have been lost. The proximal segments form a membrane which is covered with deciduous hairs. The female loses wings, legs and halteres, and burrows into the skin of the bat, so that only the posterior segments project from the skin. These segments swell and form a globular knob on which the spiracles, anus and genital opening are situated. The membrane of the anterior segments grows forward and envelops thorax and head completely (Fig. 196).

82

The abdomen of the free-living male is segmented and also has only three pairs of spiracles in segments 5 7. The male genitalia of *Ascodipteron* are not retracted into the abdomen at rest. Aedeagus and postgonites resemble those of the Nycteriboscinae in general, but the aedeagus is shorter and broader and the postgonites are symmetrical. The praegonites are large and bear spines at the apex. They were named claspers in the past.

As far as is known, the puparia of the Nycteriboscinae are attached to the substrate near the roosting places of bats. They are flattened-ovoid and have two small, round, anterior and two small posterior spiracles (Fig. 162). The puparia of *Ascodipteron* are free, ovoid and only slightly flattened dorso-ventrally and are dropped to the ground as in most Hippoboscidae. They have four more or less long, sinuate, slit-like spiracles at the posterior end (Fig. 213).

Fig. 162: *Raymondia seminuda* Jobling.
Puparium

DISTRIBUTION

There are three subfamilies with twenty-three genera and ninety-four species in America, and two subfamilies with four genera and about sixty-five species in the Old World. The Streblidae are mainly distributed in the tropics. Only a few species reach the southern parts of the Palaearctic region. Distribution is restricted in the north by the January isotherm of 8°C, as Streblidae apparently cannot survive on hibernating bats and can only exist where bats are more or less active throughout the year. The genus *Brachytarsina* is mainly Oriental (twenty-three species); only five or six species occur in the Ethiopian region and Malagasy, and a subspecies of an Ethiopian species occurs in the Middle East. One species of *Brachytarsina* occurs only in the Mediterranean region, and an Indian species occurs in Afghanistan. *Raymondia* has about fifteen species in the Ethiopian region, and only four or five species have been described from the Oriental region. An Ethiopian species occurs in the E. Mediterranean and an Indian species occurs in Afghanistan. Eleven species of *Ascodipteron* occur in Africa and two have been recorded only from the Middle East. The Oriental species of *Ascodipteron* have not been fully worked out, and there are a number of undescribed species.

HOST-PARASITE RELATIONSHIPS

The Streblidae resemble the two other families in having species-specific species, species which occur on related genera of a family and unspecific species which occur on species of different families of bats. No species are known which occur normally on both Megachiroptera and Microchiroptera. Only a few species of *Brachytarsina* occur regularly on Megachiroptera.

SYSTEMATIC PART

DIAGNOSIS OF THE FAMILY STREBLIDAE

Pupiparous, ectoparasitic, bloodsucking parasites of bats. Small or medium-sized, 1.5–5.0 mm. Colour yellowish to brown.

Head: Rounded, or triangular in profile or flattened dorso-ventrally. Head either freely movable or closely attached to thorax, with a ventral ctenidium in some American genera. Eyes with several facets, a single facet, or eyes absent. Palps 1-segmented, projecting anteriorly, or curved dorsally. Antennae 2-segmented, scape fused with head capsule. Third segment either free or partly invaginated in the second segment, which has a dorsal slit. Arista branched.

Thorax: More or less spherical or dorso-ventrally or laterally compressed. Transverse mesonotal suture complete or indistinct in the middle. Longitudinal median suture present or absent. *Wings*: Normally developed, with six longitudinal veins and three cross veins in the American genera with functional wings, with five longitudinal veins and two cross veins in the Nycteriboscinae; only two well marked longitudinal veins and no cross veins in the Ascodipterinae. *Legs*: Either normal or considerably legthened in some American genera. Claws simple. Pulvilli present. Empodium feathered or bare.

Abdomen: Membranous in its greater part, only the basal tergites and sternites and posterior end of abdomen sclerotized. Dorsum of abdomen with a bare median space bordered with long setae in the Nycteriboscinae. Wings folded longitudinally on the bare median space at rest. This bare space is less developed in the American genera with normal wings and absent in genera with reduced wings or without wings.

Male Genitalia: Male genitalia retracted into the abdomen at rest in the Nycteriboscinae, not retracted in the Ascodipterinae. Aedeagus with a very short or long connecting membrane, which bears spines in some genera. External genitalia reduced to two small praegonites (digitiform processes) in the Nycteriboscinae, larger and with spines at the apex in the Ascodipterinae.

Two subfamilies with four genera in the Old World; three subfamilies with twenty-three genera and about ninety-four species in America.

84

Streblidae: Brachytarsina

Key to the Old World Subfamilies of Streblidae

1. Sexes similar. Head normally developed. Wings with five well sclerotized longitudinal veins and two cross veins. Anal veins reduced or absent. Legs strong and thick. Abdominal spiracles 2–7 present. **Nycteriboscinae** Speiser
 – Sexes markedly different. Male free-living, of ordinary habitus of Streblidae, with small rounded head and segmented abdomen. Female with head capsule much reduced, consisting of a few sclerites separated by wide membranous areas. Theca of labium very large, conical, strongly sclerotized, with fourteen rows of strong teeth on evertible arcs on the labella. Wings with two well-marked longitudinal veins with setae. R_2+_3 reduced, other veins indicated by folds in the membrane. Cross veins absent. Only abdominal spiracles 5–7 present. The female sheds wings and legs, burrows into the skin of the bat and becomes flask-shaped by growth of the anterior abdominal segments, which enclose head and thorax. **Ascodipterinae** Monticelli

Subfamily NYCTERIBOSCINAE Speiser, 1900
Arch. Naturgesch., 66 : 31.

Key to the Palaearctic Genera of Nycteriboscinae

1. Medium-sized insects, 2.5–5.0 mm. Head rounded or rounded-triangular in profile, funnel-shaped, without ventral depressions for the fore coxae. Postvertex extending between the laterovertices. Eyes single-faceted. Fore coxae more or less widely separated. Humeral processes absent. A short anal vein present, forming an anal cell. Alula well developed. Aedeagus with a short or long connective membrane without spines. Postgonites slightly asymmetrical.
 Brachytarsina Macquart
 – Small insects, 1.5–2.0 mm. Head flattened, quadrangular or transverse-elliptical, with ventral depressions for the fore coxae. Postvertex not extending between the laterovertices. Eyes absent. Fore coxae broadly separated by sternal plate of thorax. Humeral processes present. Anal vein much reduced or absent. Alula either well developed, reduced or absent. Aedeagus with a long connective membrane with spines. Postgonites markedly asymmetrical. **Raymondia** Frauenfeld

Genus BRACHYTARSINA Macquart, 1851
(1850) *Mém. Soc. Sci. Agric. Lille*, p. 282 (*Dipt. Exot.*, Suppl. 4, p. 307).

Nycteribosca Speiser, 1900. *Arch. Naturgesch.*, 66 : 31.

Type Species: *Brachytarsina flavipennis* Macquart, 1851.
Medium-sized insects, 2.5–5.0 mm.
Head: Funnel-shaped, rounded or rounded-triangular in profile. Eyes single-faceted. Postvertex distinct, extending between laterovertices. Arista of antenna long, branched. Palps broad or oblong.

Thorax: Almost spherical, mesonotal transverse suture complete. Humeral processes absent. Fore coxae more or less widely separated anterior to sternal plate of thorax. *Wings*: With five longitudinal veins, humeral vein and two cross veins. Anal vein short, forming an anal cell. Basitarsus only slightly longer than the other tarsal segments.

Male Genitalia: Aedeagus long and slender, pointed, with a long apodeme. Connective membrane of aedeagus either very short or long, without spines. Parameres slightly asymmetrical. Anal sclerite rod-shaped, curved.

About thirty species in the tropics, a few species in the southern parts of the Palaearctic region. Two species in Israel.

Key to the Species of Brachytarsina in Israel

1. Theca of labium as long as wide or slightly longer than wide. R_1 and R_2+_3 gently curved towards the costa. R_4+_5 and M_1+_2 diverging near the wing margin. Setae on thorax dense and short. Scutellum with about sixty setae. Postnotum without conical process. Lateral processes of tergite $1+2$ triangular, with two or three apical setae. Sternite 7 of female large, rounded-triangular. Tergite 10 with straight anterior margin and a row of four long setae. Sternite 10 with sclerotized horns anteriorly. Right postgonite of male with distinctly hooked end. **B. flavipennis** Macquart

 – Theca of labium conical, much longer than wide. R_1 and R_2+_3 sharply curved towards costa. R_4+_5 and M_1+_2 parallel near the wing margin. Setae on mesonotum long, more widely spaced than in *B. flavipennis*. Scutellum with about thirty setae. Postnotum with conical process. Lateral posterior processes of tergite $1+2$ of abdomen rounded, with about five apical setae. Sternite 7 of female small, tergite 10 with a row of six to eight setae and rounded anterior margin. Sternite 10 without anterior horns. Postgonites of male with rounded ends. **B. alluaudi minor** Theodor

Brachytarsina flavipennis Macquart, 1851
Figures 158, 160, 163–166, 169, 171–176

Brachytarsina flavipennis Macquart, 1851 (1850) *Mém. Soc. Sci. Agric. Lille*, p. 282 (=*Dipt. Exot.*, Suppl. 4, p. 308).
Nycteribosca africana Walker. Jobling (1934) *Parasitology*, 26 : 86 (pro parte); (1939) *ibid.*, 31 : 150.
Nycteribosca kollari Frauenfeld, 1855, *Sber. Akad. Wiss. Wien*, 18 : 329.
Nycteribosca kollari Frauenfeld. Theodor & Moscona (1954) *Parasitology*, 44 : 227; Theodor (1954) in Lindner, 66b, Streblidae, p. 6.

Length 2.5–2.8 mm. Wing length 2.5–2.7 mm.
Head: Eyes small. Postvertex broad, nearly triangular, with eight to sixteen setae. Theca of labium as long as wide or slightly longer than wide, with rounded sides. Palps flattened, oval.
Thorax: Wider than long. Mesonotum with numerous dense, short setae. Scutellum broadly rounded, with about sixty setae and a row of setae at the posterior

margin. *Wings*: R_1 and R_2+_3 gently curved towards the costa. R_4+_5 and M_1+_2 diverging near the wing margin. Postnotum without conical process near the base of the haltere.

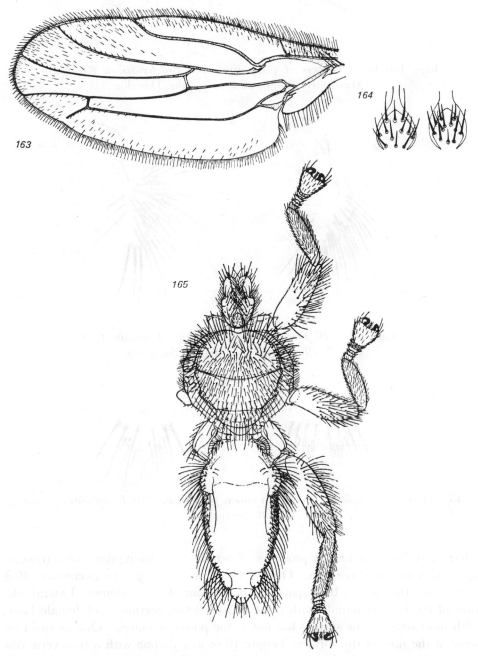

Figs. 163–165: *Brachytarsina flavipennis* Macquart. 163. wing; 164. postvertex, different forms; 165. female, dorsal (after Jobling, 1934, Parasitology, 26)

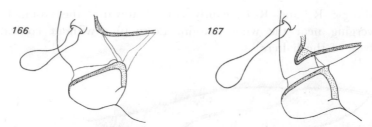

Figs. 166–167: Process of postnotum. 166. *Brachytarsina flavipennis* Macquart; 167. *B. alluaudi minor* Theodor

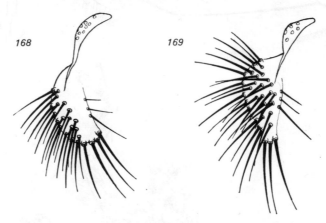

Figs. 168–169: Posterior dorsal process of tergite 1+2. 168. *B. alluaudi minor;* 169. *B. flavipennis*

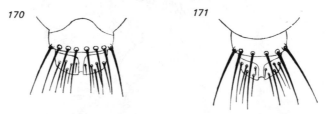

Figs. 170–171: Apical cone of abdomen of female. 170. *B. alluaudi minor;* 171. *B. flavipennis*

Abdomen: Posterior lateral processes of tergite 1+2 triangular, with truncate apex with two or three setae. One or two rows of twenty to twenty-five long setae along the median bare space on the dorsum of the abdomen. Lateral sclerites of tergite 7 of female with three or four setae. Sternite 7 of female large, with short setae on the surface but not at the posterior margin. One or two long setae at the sides of the surface. Tergite 10 of apical cone with a transverse row of four long setae and with straight anterior margin. Sternite 10 with sclerotized horns anteriorly.

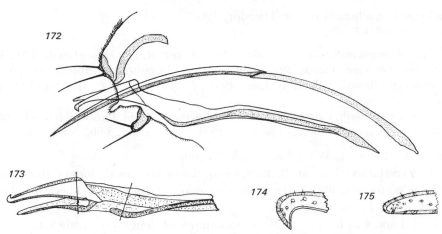

Figs. 172–175: *Brachytarsina flavipennis* Macquart. Male genitalia.
172. side view; 173. dorsal; 174. end of right postgonite;
175. end of left postgonite

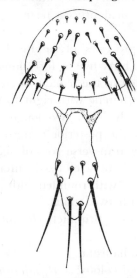

Fig. 176: *B. flavipennis*. Female, sternite 7 and 10

Male Genitalia: Right paramere distinctly hooked at the end. Hypandrium broad. Praegonites slightly conical, with two setae.

Hosts: Mainly species of *Rhinolophus*, rarely species of *Myotis, Miniopterus* and other genera.

Distribution: Mediterranean, south of January isotherm of 8°C; N. Africa; Afghanistan.

Israel: Galilee, Coastal Plain, common on *R. ferrum equinum* Schreber, *R. hipposideros minimus* Heuglin, *R. blasii* Peters, *R. euryale judaicus* And. and Matsch., rare on *Miniopterus schreibersi* Kuhl, but collection was mainly made on hibernating bats; November to April.

Brachytarsina alluaudi minor Theodor, 1968
Figures 159, 167–168, 170, 177–184

Brachytarsina alluaudi minor Theodor, 1968. *Trans. R. Ent. Soc. Lond.,* 120 : 314.
Nycteribosca alluaudi Falcoz, 1923. Jobling (1934) Parasitology, 26 : 93 (pro parte).
Nycteribosca alluaudi Falcoz. Jobling (1954) *Revue Zool. Bot. Afr.,* 50 : 102 (pro parte).
Nycteribosca alluaudi Falcoz. Theodor & Moscona (1954) *Parasitology,* 44 : 229.
Nycteribosca alluaudi Falcoz. Theodor (1954) in Lindner, 66b, Streblidae, p. 5.

Length 2.5–2.8 mm. Wing length 2.5–2.7 mm.
Head: Eyes larger than in *B. flavipennis*. Postvertex oval, longer than wide, narrower posteriorly, with three to nine setae in the distal part. Theca of labium conical, much longer than wide.
Thorax: Thorax as long as wide. Mesonotum with long and strong setae which are more widely spaced than in *B. flavipennis*. Scutellum with about thirty setae on the surface, but not on the posterior margin. *Wings*: R_1 and R_{2+3} sharply curved towards the costa. R_{4+5} and M_{1+2} parallel near the wing margin. Postnotum with a conical process near the base of the haltere. Alula rounded distally.
Abdomen: Posterior lateral processes of tergite 1+2 rounded apically, with five or six setae at the apex. Thirty-five to forty long setae along the dorsal median bare space of the abdomen in the female, twenty-five to thirty in the male. Lateral sclerites of tergite 7 of female with eight to ten long setae. Sternite 7 of female small, rounded anteriorly, with two longer and several short setae on the surface and short setae at the posterior margin. Tergite 10 of female with convex anterior margin and a transverse row of six to eight setae. Sternite 10 slightly widened anteriorly, without sclerotized anterior horns.
Male Genitalia: Both parameres with rounded ends. Hypandrium narrow. Praegonites slender, cylindrical, with two setae.
Hosts: *Rhinopoma microphyllum* Brünnich, *R. hardwickei arabium* Thomas, *Taphozous perforatus* Geoffroy.
Distribution: Egypt; Iran; Afghanistan.
Israel: Tiberias (7); in a mixed colony of *R. microphyllum* and *R. hardwickei*; January to March. Numerous specimens.

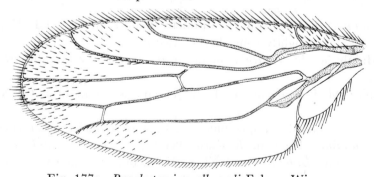

Fig. 177: *Brachytarsina alluaudi* Falcoz. Wing

Fig. 178: *B. alluaudi minor* Theodor. Postvertex, different forms

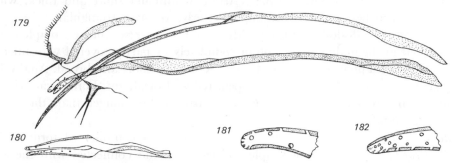

Figs. 179–182: *B. alluaudi minor*. Male genitalia. 179. side view; 180. dorsal; 181. end of right postgonite; 182. end of left postgonite

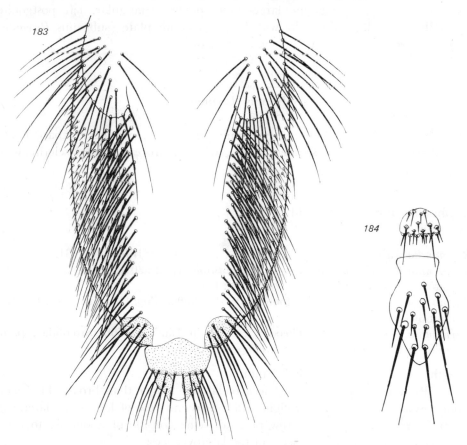

Figs. 183–184: *Brachytarsina alluaudi minor* Theodor. Female. 183. abdomen, dorsal; 184. sternites 7 and 10

Diptera Pupipara

Genus RAYMONDIA Frauenfeld, 1855
Sber. Akad. Wiss. Wien, 18 : 328.

Type Species: *Raymondia huberi* Frauenfeld, 1855.
Small insects, 1.5–2.0 mm.
Head: Flattened, with ventral depressions for the fore coxae. Postvertex not extending between the laterovertices. Arista of antenna short and thick, with numerous branches. Palps extending anteriorly, truncate-triangular.
Thorax: Depressed dorso-ventrally. Mesonotal transverse suture complete or interrupted in the middle. Mesonotum completely or partly covered with microtrichia. Humeral processes present. *Wings*: Relatively short and broad. Anal vein very short or absent. Alula well developed, reduced or absent. Fore coxae widely separated by sternal plate of thorax. Basitarsus slightly longer than the other tarsal segments.
Male Genitalia: Aedeagus either short and broad, of complicated form (subgenus *Brachyotheca* Maa) or simple and elongate and connected with the abdomen by a long connective membrane with spines (subgenus *Raymondia*). Apodeme of aedeagus either long or short. Postgonites (parameres) markedly asymmetrical, right postgonite large, more or less triangular, left postgonite markedly reduced. Anal sclerite with broad posterior plate (subgenus *Raymondia*) or narrowly triangular (subgenus *Brachyotheca*).

Subgenus **Raymondia**

Setae on sternal plate of thorax uniformly distributed, not leaving bare longitudinal stripes. Aedeagus slender, simple, without processes. Anal sclerite with broad posterior plate.

Raymondia huberi Frauenfeld, 1855
Figures 161, 185–195

Raymondia huberi Frauenfeld, 1855. *Sber. Akad. Wiss. Wien,* 18 : 331.
Raymondia huberi setosa Jobling, 1930. *Parasitology,* 22 : 301.
Raymondia setosa Jobling, 1939. *Ibid.,* 31 : 154.
Raymondia huberi Frauenfeld. Jobling (1954). *Revue Zool. Bot. Afr.,* 50 : 104 (redescription).
Raymondia setosa Jobling. Theodor (1954) in Lindner, 66b, Streblidae, p. 8; Theodor & Moscona (1954) *Parasitology,* 44 : 230.

Length 1.7 mm. Wing length 1.3 mm.
Head: Rounded-quadrangular. Theca of labium longer than broad. The main gular row consists of seven or eight setae; a second row of long setae lateral to the main row. The posterior row of small setae consists of about six to eight setae which do not reach the setae of the laterovertices.

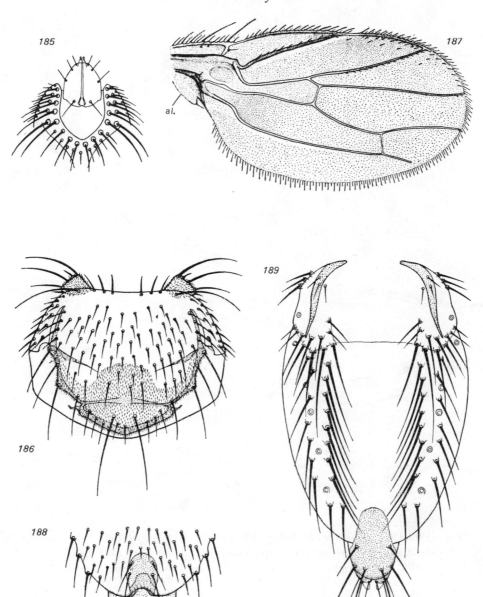

Figs. 185–189: *Raymondia huberi* Frauenfeld. 185. head, ventral;
186. thorax, dorsal; 187. wing (after Jobling, 1930, *Parasitology*,
22:300); 188. female abdomen, ventral, posterior part;
189. female abdomen, dorsal

al. – alula

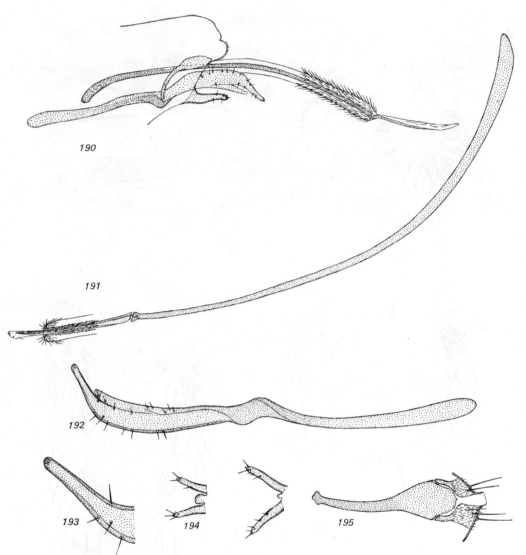

Figs. 190–195: *Raymondia huberi* Frauenfeld. Male genitalia. 190. genitalia extended; 191. aedeagus and apodeme; 192. postgonites and apodeme; 193. end of right postgonite; 194. praegonites; 195. anal sclerite

Thorax: The setae on the mesonotum differ in length in male and female, being markedly longer in the male. The setae reach to the anterior margin of the mesonotum, but are less numerous anteriorly. Only two or three setae behind the lateral parts of the transverse suture. Microtrichia form a broad stripe before the scutellar suture. Sternal plate covered to the anterior margin with microtrichia, which form distinct, dense, transverse rows.

Abdomen: Two rows of about ten long setae border the median bare space on the dorsum. Tergite 10 of female triangular anteriorly, with rounded apex and with a transverse row of four long setae posteriorly and two lateral setae anteriorly; the sclerite reaches distinctly beyond the anterior setae.

94

Genitalia: Aedeagus of male straight, relatively short, apodeme very long, three or four times longer than the aedeagus. Spines on connective membrane thin and short. Right postgonite large, curved, with a long apical process with rounded end, seven or eight setae at the outer margin and a few at the inner margin. Praegonites long and slender with a few minute hairs in the apical half. Anal sclerite long, with oblong end-plate.

Hosts: *Cardioderma cor* Peters, *Hipposideros commersoni* Geoffroy, *Triaenops persicus* Dobson, *Asellia tridens* Geoffroy.

Distribution: Afghanistan; Arabia; Egypt; E. Africa.

Israel: Jaffa (13), one female on *Asellia tridens*; December 1949.

Subfamily ASCODIPTERINAE Monticelli, 1898
Ric. Lab. Anat. Norm. R. Univ. Roma, 6 : 255 (as family):

Diagnosis as for the genus *Ascodipteron*.

Genus ASCODIPTERON Adensamer, 1896
Sber. Akad. Wiss. Wien, 10 : 400.

Type Species: *Ascodipteron phyllorhinae* Adensamer, 1896.

Sexes markedly different. Male with small, rounded head and small, transversely elliptical theca of labium, free-living, of ordinary habitus of Streblidae. Eyes absent. Palps present, 1-segmented. Wings with R_1 and R_{4+5} sclerotized and with setae, R_{2+3} more or less reduced, without setae; other veins indicated by folds in the membrane. Cross veins absent. Legs long and slender. Basitarsus long. Abdomen segmented, but with only three pairs of spiracles in segments 5–7. Genitalia not retracted inside the abdomen at rest. Aedeagus long, pointed, with a connective membrane without spines and with a long apodeme. Postgonites (parameres) symmetrical, with long apodeme. Praegonites (claspers) with several short, curved spines at the end.

Female with sclerites of head strongly reduced, connected by wide membranous areas. Eyes absent. Genae triangular. A small occipital sclerite, a frontal sclerite which bears the antennae and two truncate-triangular laterovertices on the dorsal surface. Theca of labium very large, truncate-conical, strongly sclerotized, with fourteen rows of strong, curved, blade-like teeth on evertible arcs on the labella. Palps absent. Only three pairs of abdominal spiracles concentrated at end of abdomen. Segmentation of abdomen obliterated. Legs and wings of newly hatched female as in the male.

About sixteen species, mainly in Africa. Two species in the Middle East and several species in the Oriental region. Two species in Israel.

95

Diptera Pupipara

Ascodipteron rhinopomatos Jobling, 1952
Figures 196–203, 207–210, 213

Ascodipteron rhinopomatos Jobling, 1952. *Parasitology*, 42 : 126.
Ascodipteron rhinopomatos Jobling, 1952. Theodor (1954) in Lindner, 66b, Streblidae, p. 9; Theodor & Moscona (1954) *Parasitology*, 44 : 231 (description of male and winged female).

Embedded Female: Length 3.5–4.0 mm.
Head: Length of theca of labium 0.40–0.48 mm, width posteriorly 0.30–0.32 mm. Four to five rows of unpigmented pegs, which form a triangular area at the sides of the dorsal surface of the theca. Teeth of the four median dorsal rows of labellum very thin, S-shaped, with secondary serrations. Teeth of lateral and ventral rows blade-shaped. Laterovertices narrow, truncate-triangular. Occipital sclerite with two anterior triangular processes. Genae narrow, triangular, with fifteen to twenty-eight unpigmented pegs.
Thorax: Scutellum with a seta at each side. Mesonotum more or less rectangular. Episternum and epimeron of mesopleuron with fine setae and with pegs. Fore coxae curved, with fine setae. Hind coxae longer than wide.
Abdomen: Dorsal row of spiracles at posterior end slightly curved; an irregular row of setae between the dorsal spiracles and the cerci. Cerci small, of irregular form, not projecting from the surface, with two or three longer and several shorter setae. Setae on posterior part of abdomen long and thin, the two proximal rows of spines on segment 5 consist of short spines, which are not thicker than the posterior setae.
Alate Female: Length 1.6–1.8 mm. Wing length 1.65 mm, width 0.8 mm, with an indentation at the margin at end of M_1+_2. Only R_1 and R_4+_5 well sclerotized and bearing setae. R_2+_3 reduced in length, but distinct and without setae. Anterior four segments of abdomen cylindrical, with deciduous hairs. Limits of segments obliterated. Segments 5–7 indicated by rows of setae, with spiracles. Halteres large, slender. Legs slender, basitarsus as long as tarsal segments 2–4 together. Wings kept flat above the abdomen, not folded longitudinally as in the Nycteriboscinae.
Male: Length 1.4–1.5 mm. Wing length 1.4–1.5 mm, width 0.7 mm.
Weakly sclerotized. Head small, rounded in profile, rectangular in dorsal view. Laterovertices elliptical, separated from the postgenae. Theca of labium transversely elliptical. Palps narrow, tapering to a point, with a short seta. Mesonotum as in the female, but narrower anteriorly. Wings as in the female. Abdomen cylindrical, segmented, also with only three pairs of spiracles on segments 5–7.
Genitalia: Aedeagus straight, pointed, 0.28 mm long, its apodeme about as long as the aedeagus. Postgonites triangular, with rounded apex and with a long apodeme. Praegonites slender, with two or three short, curved spines at the end.

96

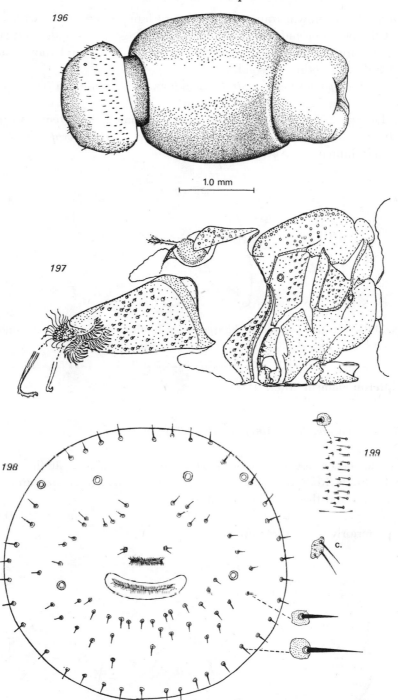

1.0 mm

Figs. 196–199: *Ascodipteron rhinopomatos* Jobling. Female. 196. embedded
female; 197. head and thorax, lateral; 198. posterior end of abdomen,
posterior view; 199. proximal rows of spines, lateral
c. – cercus

Puparium: Light brown, transparent, 1.5 mm long, 1 mm wide, with four spiracles at the posterior end, which are about twice as long as wide, and two non-functional spiracles. The puparium is dropped to the ground, not attached to the substrate as in other Streblidae.

Hosts: *Rhinopoma hardwickei arabium* Thomas, *R. microphyllum* Brünnich.

Distribution: Afghanistan; Egypt; Somalia.

Israel: Tiberias (7), about twenty embedded females, three females and three males bred from puparia, from a mixed colony of *R. microphyllum* and *R. hardwickei*; January to April.

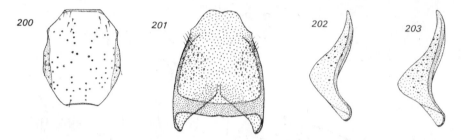

Figs. 200–203: *Ascodipteron rhinopomatos* Jobling. Female. 200. mesonotum; 201. theca, dorsal; 202–203. gena

Ascodipteron namrui Maa, 1965
Figures 204–206, 211–212

Ascodipteron namrui Maa, 1965. *J. Med. Ent.*, 1 : 313.

This species occurs together with *A. rhinopomatos* and was at first not recognized. It differs from *A. rhinopomatos* mainly in its larger size and different form of gena and theca of the labium.

Embedded Female: Length 4–5 mm. Theca of labium: length 0.47–0.55 mm, width posteriorly 0.40–0.45 mm.

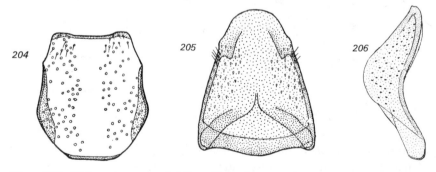

Figs. 204–206: *A. namrui* Maa. Female. 204. mesonotum; 205. theca;
206. gena

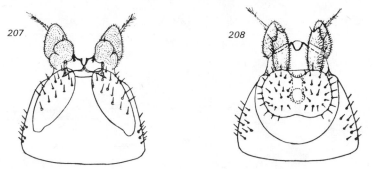

Figs. 207–208:   *A. rhinopomatos.* Male head. 207. dorsal; 208. ventral

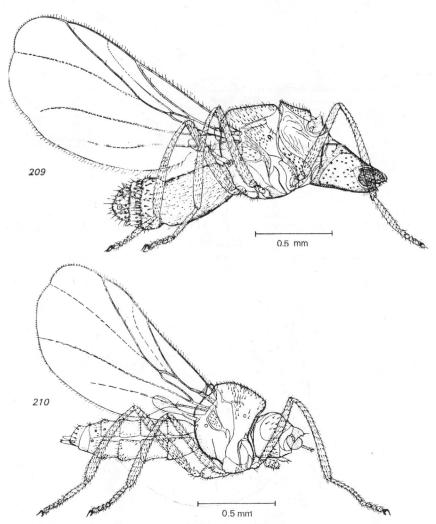

Figs. 209–210:   *Ascodipteron rhinopomatos* Jobling. 209. young female; 210. male

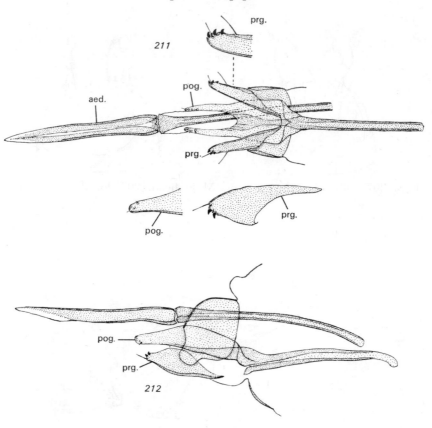

Figs. 211–212: *Ascodipteron namrui* Maa. Male genitalia. 211. ventral;
212. lateral

aed. – aedeagus; pog. – end of postgonite, lateral; prg. – praegonite

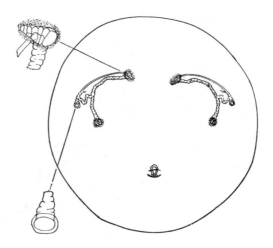

Fig. 213: *Ascodipteron rhinopomatos* Jobling. Puparium, spiracles

Gena broader than in *A. rhinopomatos*, with rounded dorsal apex, with forty to fifty-five pigmented pegs. Mesonotum broader. Legs of alate female stronger, femora and tibiae broader than in *A. rhinopomatos*. Wings 2 mm long, 1 mm wide, with rounded apex, without indentation at the end of M_1+_2.

Male: Resembling that of *A. rhinopomatos,* but palps rounded or truncate anteriorly. Genitalia as in *A. rhinopomatos,* but aedeagues 0.38 mm long and relatively broader than in *A. rhinopomatos*. Praegonites broadly triangular.

Puparium: 1.8 mm long, 1.2 mm wide, otherwise as in *A. rhinopomatos*.

Hosts: Only *Rhinopoma microphyllum* in Egypt.

Distribution: Egypt.

Israel: Tiberias (7), four embedded females, two males and two females bred from puparia, from a mixed colony of *R. microphyllum* and *R. hardwickei*; January to April.

Family NYCTERIBIIDAE Westwood, 1840
Introduction to the Modern Classification of Insects, II, p. 154.

INTRODUCTION

The Nycteribiidae are pupiparous, blood-sucking, obligatory ectoparasites of bats. Their biology has resulted in far-reaching adaptations which in some respects have resulted in developments opposite to those in other Diptera. Thus, the sternites of the thorax have become fused into a broad plate, while the mesonotum is membranous. The pleurae are displaced dorsally and the legs are inserted on the dorsal surface, giving the insect a spider-like appearance (Fig. 214). The head is kept folded back at rest, so that its dorsal surface rests on the mesonotum. It is rotated forward through 180° for feeding. Wings are absent in all Nycteribiidae, but halteres are present.

MORPHOLOGY

Head: The head (Figs. 221–226) is rounded or laterally or dorso-ventrally compressed. The head capsule is formed by a helmet-shaped sclerite which covers the dorsal and lateral surfaces of the head. Its anterior and ventral surfaces are membranous. The dorsal surface is also membranous anteriorly in some species. The vertex usually bears setae in the anterior part. Eyes present or absent. They consist of a single facet, two more or less separated facets in some genera, and one species has two ocular frames on each side, each with two facets. Antennae (Figs. 227–228) 2-segmented, scape fused with the head capsule. The third segment (apparent second) is completely invaginated inside the basal segment, so that only the dendriform arista (Fig. 229) is visible outside. Basal segment with a dorsal slit and with a dorsal process which differs in form in the various genera. Maxillary palps (Figs. 230–233) 1-segmented flattened, varying in form and with setae at the margin, also on the ventral side in some genera and usually with a long or very long terminal seta. The proboscis consists of a broad, usually conical, part, the theca, and a narrow anterior part, the labella, the length of which varies widely in different genera. The structure of head and mouth parts has been described in detail by Jobling (1928).

Thorax: The thorax (Figs. 234–235) has been so greatly modified that the original limits of the segments are obliterated to a great extent. It has been studied in detail by Schlein (1970). Basisterna 2 and 3 form a broad plate which is divided by a median suture and by two oblique sutures between basisterna 2 and 3. The anterior margin of basisternum 2 is raised to facilitate passage between the hairs of the host. The suture between basisterna 2 and 3, the meso-metasternal suture, is fused in some genera and is then termed 'closed', or

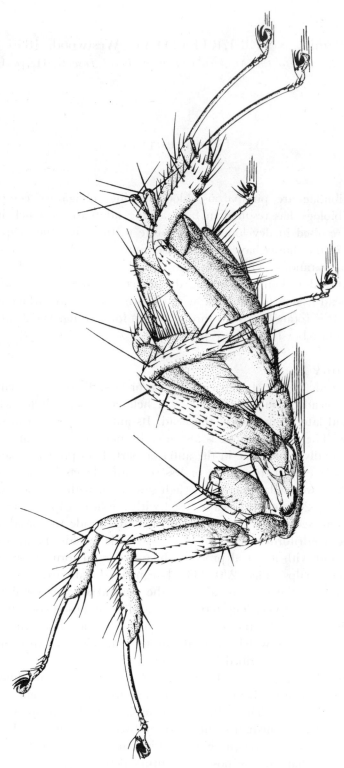

Fig. 214: *Stylidia biarticulata* (Hermann). Female

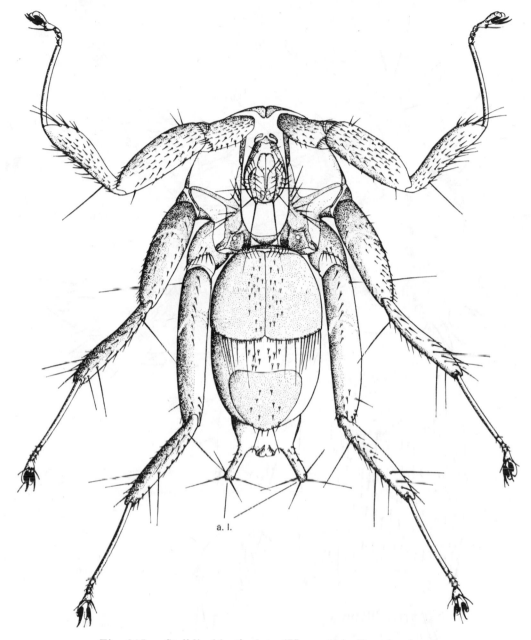

Fig. 215: *Stylidia biarticulata* (Hermann). Female, dorsal
a.l. – anal lobes

it forms a membranous stripe and is then termed 'open'. The prothorax is reduced to a few small sclerites and has been displaced to the dorsal surface around
the insertion of the head. The dorsal surface is membranous in its greater part.
Two longitudinal ridges, the notopleural sutures, enclose a median field, the

105

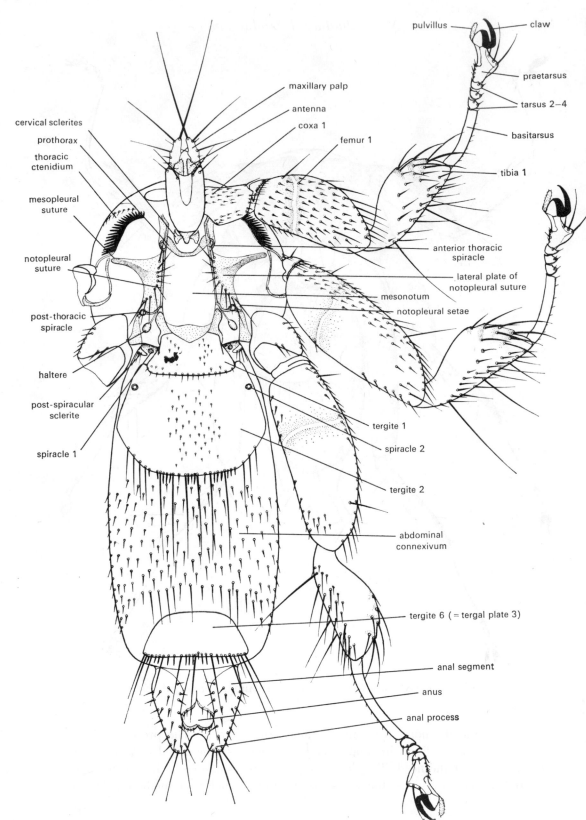

Fig. 216: *Nycteribia latreillii* (Leach). Female, dorsal

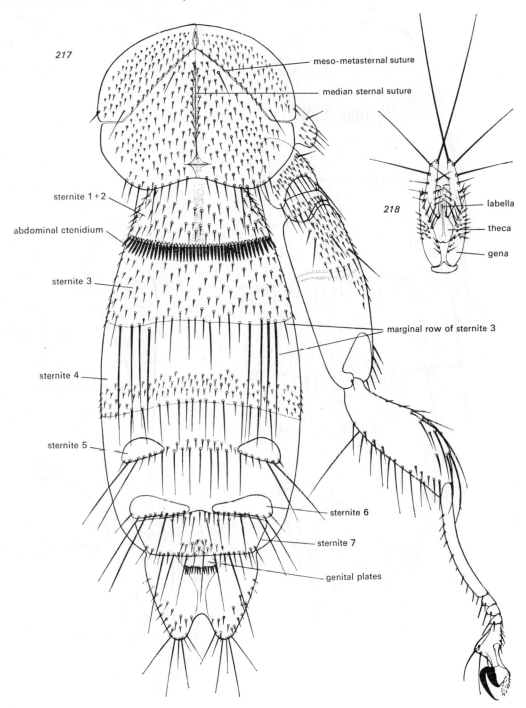

217

meso-metasternal suture

median sternal suture

218

labella

theca

gena

sternite 1+2

abdominal ctenidium

sternite 3

marginal row of sternite 3

sternite 4

sternite 5

sternite 6

sternite 7

genital plates

Figs. 217–218: *Nycteribia latreillii* (Leach). Female. 217. ventral;
218. head, ventral

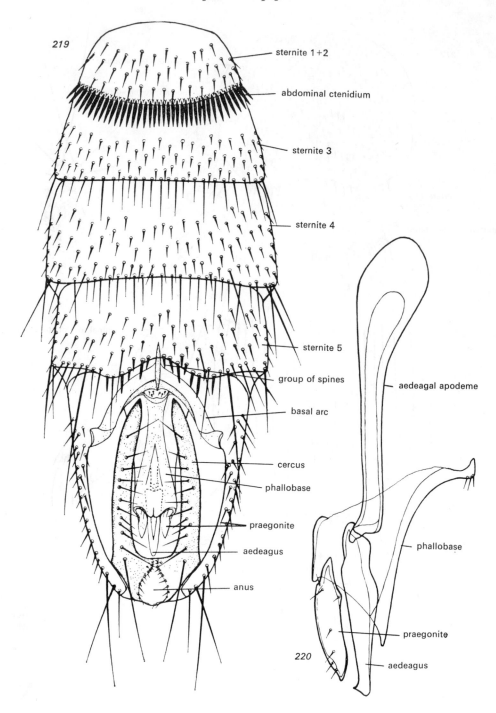

Figs. 219–220: *Nycteribia latreillii* (Leach). Male. 219. abdomen, ventral;
220. genitalia

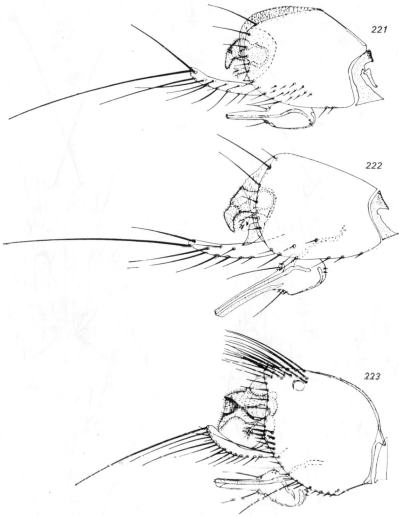

Figs. 221–223: Head, side view. 221. *Nycteribia latreillii* (Leach);
222. *N. vexata* Westwood; 223. *Penicillidia dufourii* (Westwood)

mesonotum. The anterior spiracles are situated inside them, more or less anteriorly. The posterior spiracles are situated on a branch of the notopleural sutures
close to the haltere groove. Sclerotized plates of varying form adjoin the notopleural sutures laterally. These lateral plates bear a characteristic row of setae,
the notopleural setae, which are absent only in a few species and give important
systematic characters. The areas lateral to the notopleural sutures represent the
pleurae. The mesopleural suture is a more strongly sclerotized ridge from the
anterior part of the lateral plates to the anterior aspect of coxa 2. The posterior
end of the mesonotum is limited by a vertical plate which sometimes bears a
process, the postnotum. The halteres have a thin stalk and a spherical or ovoid
head. The haltere groove may be open, or closed completely or partly, by a
sclerotized flap, the cover of the haltere groove. At the anterior lateral parts of
the thorax, between coxae 1 and 2, is a movable sclerite which bears a row of

109

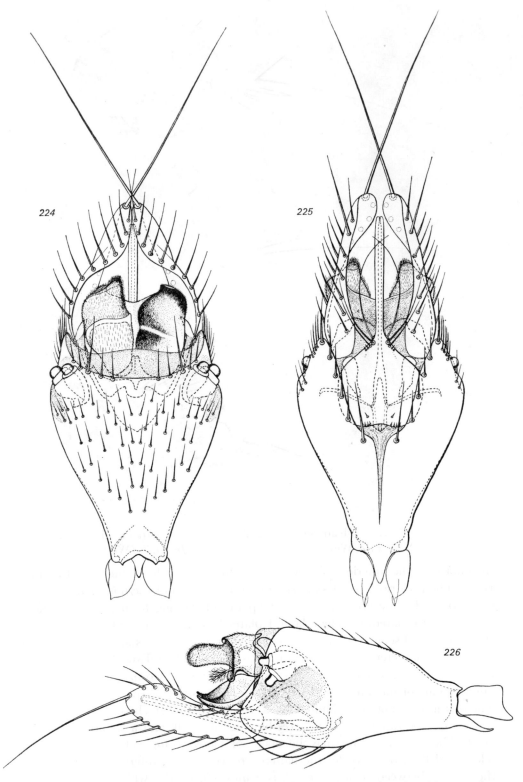

Figs. 224–226: *Cyclopodia sykesii* (Westwood). Head. 224. dorsal; 225. ventral; 226. head, lateral (after Jobling, 1928, *Parasitology*, 20)

strong, curved spines, the thoracic ctenidium. This is a differentiation of coxa 2 which has assumed a new function, grasping the fur of the host. This organ is a new development and is present only in the Nycteribiidae, in all genera except in the subgenus *Eremoctenia* Scott of the genus *Penicillidia* Kolenati. The fore coxae are either conical and very long in some genera, or shorter and broader in others. Coxae 2 and 3 are short, ring-shaped.

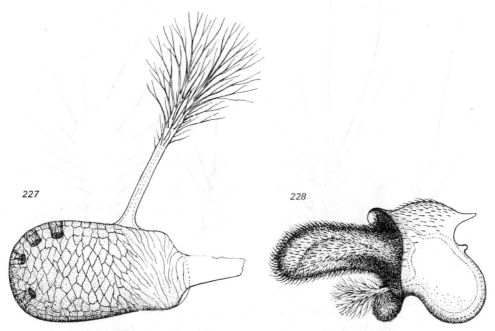

Figs. 227–228: *Cyclopodia sykesii* (Westwood). 227. antenna, 3rd segment; 228. antenna, side view (from Jobling, 1928, Parasitology, 20)

Fig. 229: *Nycteribia latreillii* (Leach). Arista

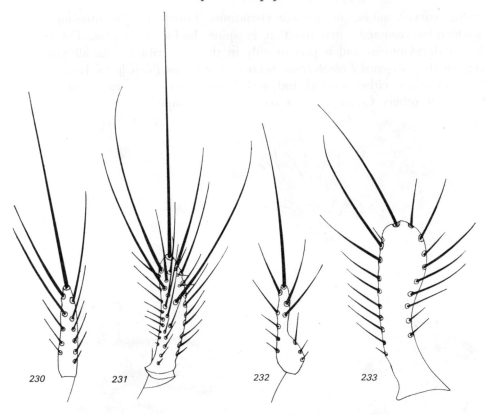

Figs. 230–233: Palps. 230. *Nycteribia latreillii* (Leach); 231. *Penicillidia dufourii* (Westwood); 232. *Eucampsipoda hyrtlii* (Kolenati); 233. *Cyclopodia sykesii* (Westwood)

Legs: The femora are thick and have a ring of weaker integument near the base. The tibiae vary in form: they are either laterally compressed and bear several rows of setae in the distal part of the ventral margin or they are more or less cylindrical, with two or three rings of weaker integument and short setae in the middle (Figs. 236–240). The basitarsus is usually very long or shorter and bears half-rings of weaker integument. All these rings of weaker integument increase flexibility. Tarsal segments 2–4 short. Praetarsus triangular, broad, with two strong claws and pulvilli, but without empodium.

Abdomen: The abdomen consists of seven segments before the anal segment. Seven pairs of spiracles are present in all species, but their position varies considerably in some species. The original number of tergites and sternites of the female is usually reduced by fusion of the sclerites or by their disappearance in connection with the membranization of the abdomen.

Male Abdomen (Figs. 219, 241-242): The male has preserved the segmentation of the abdomen to a higher degree than the female. There are six tergites before

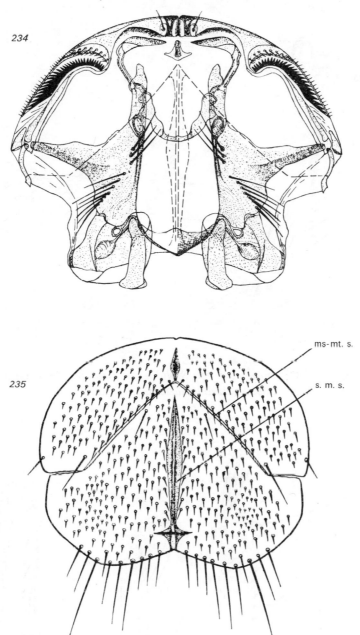

Figs. 234–235: *Nycteribia latreillii* (Leach). Thorax. 234. dorsal; 235. ventral
ms-mt.s. – meso-metasternal suture; s.m.s. – sternal median suture

the anal segment, but tergites 1 and 2 are fused in the subfamily Cyclopodiinae and in the males of some species of *Penicillidia*. Sternite 1 has disappeared or is fused with sternite 2. A small strip-like sclerite, the postspiracular sclerite, lies

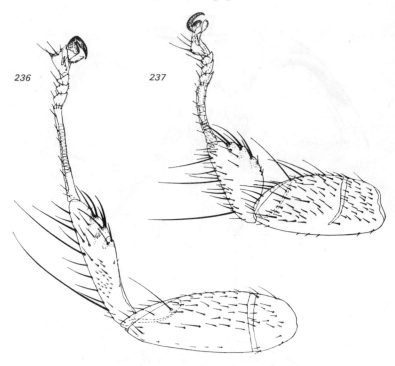

Figs. 236–237: Forelegs. 236. *Stylidia biarticulata* (Hermann);
237. *Nycteribia latreillii* (Leach)

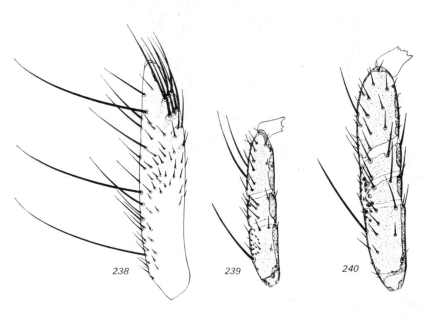

Figs. 238–240: Fore tibiae. 238. *Penicillidia dufourii* (Westwood);
239. *Eucampsipoda hyrtlii* (Kolenati); 240. *Cyclopodia greeffi* Karsch

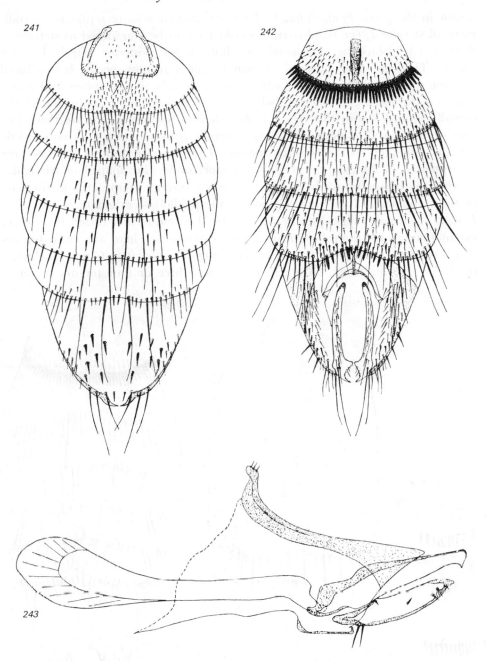

Figs. 241–243: *Nycteribia latreillii* (Leach). Male abdomen. 241. dorsal; 242. ventral; 243. genitalia, lateral

at the base of the abdomen, near the base of the basal sternite at each side (Figs. 246–251). It bears a row of setae at the posterior margin in some genera, a single apical seta in others and no setae in the genus *Cyclopodia* Kolenati. It is

115

absent in the genus *Penicillidia*. If the postspiracular sclerite represents a rudiment of sternite 1, the basal sternite would have to be considered as sternite 2. Sternite 1+2 bears the abdominal ctenidium at the posterior margin in most species. This consists of flattened, contiguous spines in most species. It is reduced in some species and consists of ordinary, more widely spaced spines and is absent in some species of genera in which there is a tendency to reduction of the ctenidium, and also in the genus *Archinycteribia* Speiser. There are three further sternites in all male Nycteribiidae. Sternite 5 usually bears a characteristic armature of spines at the posterior margin. The posterior sternites have become incorporated in the external genitalia. A more or less arc-shaped or triangular sclerite, the hypandrium (basal arc), forms an anterior frame around the base of the genitalia, and the muscles moving the phallobase originate on it.

Female Abdomen (Figs. 244–245): This has become membranized to a larger extent than in the male to permit greater expansion. There are seven tergites and seven sternites only in the American genus *Hershkovitzia* Guimarães, tergites 1 and 2 being fused. In other genera, two, three or four tergites or tergal

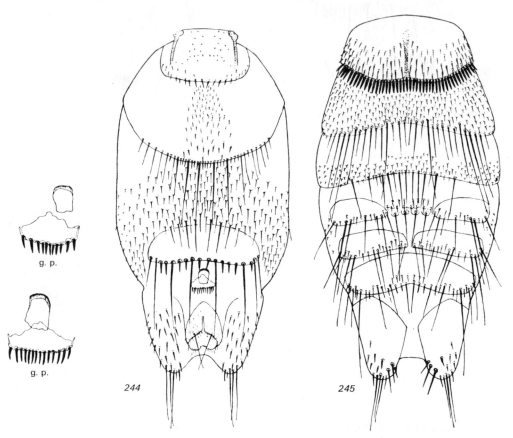

Figs. 244–245: *Nycteribia latreillii* (Leach). Female abdomen. 244. dorsal, and genital plates (g.p.); 245. ventral

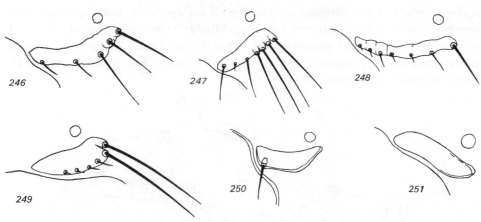

Figs. 246–251: Post-spiracular sclerites. 246. *Nycteribia latreillii* Leach;
247. *Stylidia biarticulata* (Hermann); 248. *Basilia nattereri* Kolenati;
249. *B. daganiae* Theodor and Moscona; 250. *Eucampsipoda hyrtlii*
(Kolenati); 251. *Cyclopodia greeffi* Karsch

plates are present before the anal segment, some of which may not correspond
to a single segment, and the term 'tergal plates' is used in these cases. There are
no sclerites in some genera between the basal sclerites and the anal segment, for
example, in the genus *Cyclopodia* (not in the genus *Leptocyclopodia* Theodor
which was considered as a subgenus of *Cyclopodia* in the past, but is now
regarded as a distinct genus). In other genera, tergite 6 and sternites
5 and 6 may be present. Sternite 1+2 is always present. Sternites 3 and 4 are
membranous in most species; they bear sclerites in only a few species. Sternites 5
and 6 bear sclerites which are usually divided into lateral sclerites, and sternite
7 is rarely divided. Small sclerites anterior and posterior to the genital opening,
the genital plates, probably represent rudiments of posterior sternites.

Male Genitalia: The genitalia of the Nycteribiidae differ from those of the Streblidae and Hippoboscidae in the presence of movable cerci (claspers) and in that
they are not retracted inside the abdomen at rest, except in the genus *Cyclopodia*, in which a new development has taken place, as they differ in principle
from the structure of the genitalia of the Streblidae and Hippoboscidae.

There are three types of male genitalia in the Nycteribiidae:

1. *Nycteribia type* (Figs. 243, 252). The aedeagus consists of two sclerotized
plates which are fused ventrally. It articulates with a long, plate-shaped apodeme, and lies free inside the phallobase, from which it can be protruded for a
short distance. The phallobase bears two praegonites (parameres) which form
a sheath around the aedeagus at rest and are turned laterally during copulation;
they are fused with the phallobase in some genera. The aedeagus is everted by
the action of muscles from the hypandrium to the end plate of the apodeme.
The phallobase articulates with the hypandrium, so that the aedeagus is turned
through an angle of 90°, but is protruded only for a short distance.

2. *Eucampsipoda type* (Figs. 253–255). This resembles the *Nycteribia* type

117

in the presence of a sclerotized aedeagus with an apodeme, but the aedeagus is connected with the phallobase by a long, tubular connective membrane which bears spines. The dorsal membrane of the aedeagus forms a large endophallus at the end, which varies in form in different genera. The praegonites are reduced to small, triangular sclerites.

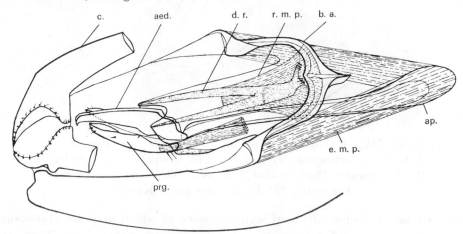

Fig. 252: *Nycteribia latreillii* (Leach). Male genitalia, semi-diagrammatic half profile showing muscle attachments

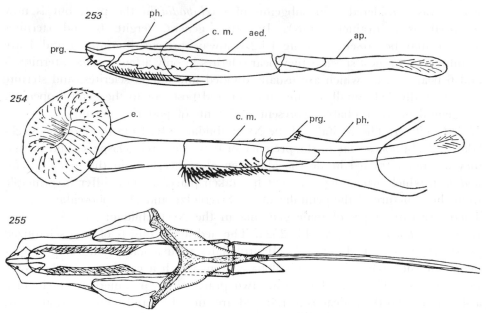

Figs. 253–255: *Eucampsipodia hyrtlii* (Kolenati). Male genitalia.
253. retracted, lateral; 254. extended; 255. ventral
aed. – aedeagus; ap. – apodeme; b.a. – basal arc; c. – cercus;
c.m. – connecting membrane; d.r. – dorsal ridge; e. – endophallus;
e.m.p. – erector muscle of phallobase; ph. – phallobase; prg. – praegonite;
r.m.p. – retractor muscle of phallobase

118

3.   *Cyclopodia* type (Figs. 256–257). The aedeagus has become membranous and is supported by a rod-like sclerite in its wall. The apodeme of the aedeagus has disappeared and a long, strip-like retractor muscle which originates on sternite 5 is inserted at the base of the aedeagus. A long, membranous connective tube, which is partly covered with spines, connects the base of the aedeagus

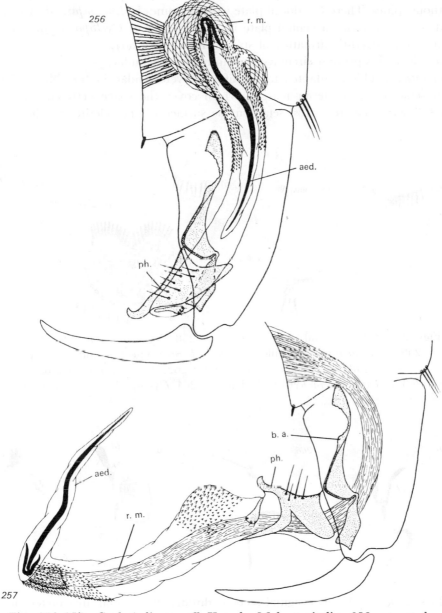

Figs. 256–257: *Cyclopodia greeffi* Karsch. Male genitalia. 256. retracted;
257. extended

aed. – aedeagus; b.a. – basal arc; ph. – phallobase; r.m. – retractor muscle

with the phallobase. The praegonites are reduced to small sclerites in the membrane. The aedeagus is retracted inside the abdomen at rest. The phallobase is usually triangular and bears an apical hook. This type of genitalia is present only in the genus *Cyclopodia*.

Female Genitalia: There are one or two genital plates in the females of the subfamily Nycteribiinae, a dorsal plate which usually bears spines and a ventral plate without spines. There is a dorsal plate without spines in *Eucampsipoda* Kolenati, and sternite 7 forms a genital plate in most species of *Cyclopodia*, in which it bears a characteristic armature of spines (Figs. 258–268).

There is a hypopygium circumversum in all Nycteribiidae.

Chaetotaxy: This is adapted to the life of Nycteribiidae in fur. Most sclerites bear setae at the posterior margin, which cover the space between them and the following sclerites. Most setae on the surface of the sclerites are horizontal

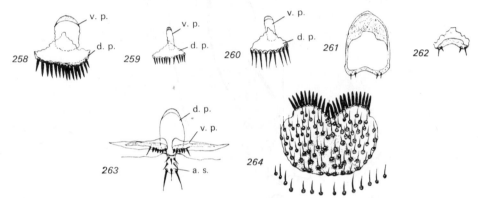

Figs. 258–264: Genital plates of females. 258: *Nycteribia latreillii* (Leach); 259. *N. pedicularia* Latreille; 260. *N. kolenatii* Theodor and Moscona; 261. *Stylidia biarticulata* (Hermann); 262. *N. vexata* Westwood; 263. *Eucampsipoda hyrtlii* (Kolenati); 264. *Cyclopodia greeffi* Karsch

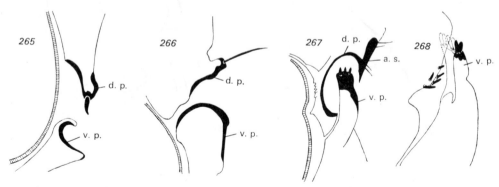

Figs. 265–268: Median sagittal sections through the genital area of females. 265. *N. latreillii;* 266. *Penicillidia dufourii* (Westwood); 267. *E. hyrtlii;* 268. *C. sykesii* (Westwood)

a.s. – anal sclerite; d.p. – dorsal plate; v.p. – ventral plate

and directed posteriorly, but some are vertical and probably have a sensory function. The number, length and position of the setae vary considerably, so that they are less constant and reliable as systematic characters than in other groups of Diptera. There are also spines, that is, short, thick, rigid setae, in a number of places. The most important spines are those forming the thoracic and abdominal ctenidia and the spines at the posterior margin of sternite 5 of the male. The notopleural setae, the row of setae at the posterior margin of the sternal plate of the thorax and the arrangement of the long setae on the ab- domen of females of the genus *Cyclopodia* are particularly important.

Puparia: These are black, shallowly convex, with two anterior and two posterior spiracles (Fig. 269). They are deposited on the substrate near the roosting places of bats and pressed to it, so that their ventral surface is flattened. The puparium has a dorsal operculum, which is opened by the action of the fore legs, which are folded above the head. There is no ptilinum.

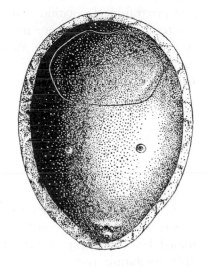

Fig. 269: Puparium of *Nycteribia* species

DISTRIBUTION

There are eleven genera of Nycteribiidae with about two hundred species which are distributed throughout the world, mainly in the tropics. The greatest num- bers of species occur in the Oriental region (about seventy) and in the Pacific region (about forty-five). Only the genus *Basilia* Miranda-Ribeiro occurs in both the Old World and in America. *Basilia* apparently entered America with vespertilionid bats. The primitive genus *Hershkovitzia* occurs only in America. Only about thirty species are distributed in the Palaearctic region, all belonging to the subfamily Nycteribiinae, except one species of *Eucampsipoda,* which entered the Middle East with a species of *Rousettus* from the Ethiopian region, and a species of *Cyclopodia* in Japan, which entered from the Oriental region with a species of *Pteropus*. The Palaearctic fauna of Nycteribiidae consists mainly of species which are restricted to the Palaearctic region, but two Ethiopian species have entered the Mediterranean and several Oriental species have spread to the E. Palaearctic. A number of Indian species occur in Afghanistan, which they entered through the western Himalayas.

The fauna of Nycteribiidae of Israel and adjacent areas consists mainly of Palaearctic (Mediterranean) species. One of them, *Stylidia biloba*, has so far been found only in Israel, but it probably has a wider distribution. One Ethiopian species, *Nycteribia schmidlii*, is common in Israel, mainly on *Miniopterus schreibersi*.

Host–Parasite Relationship

There are three degrees of host–parasite specificity, as in the Streblidae and Hippoboscidae: (a) restriction to one species of hosts; (b) restriction to a genus or several related genera of a family of hosts; (c) lack of specificity. Specificity seems to be determined in some cases, at least partly, more by ecological factors and geographical isolation than by true specificity.

Nycteribiidae do not occur normally or even feed on any hosts other than Chiroptera. The Nycteribiinae are parasites of Microchiroptera and the Cyclopodiinae of Megachiroptera. No species occurs regularly on both suborders of bats or on species of the suborder which does not contain the normal hosts of the subfamily.

Systematic Part

Diagnosis of the Family Nycteribiidae

Head: Either laterally compressed, rounded or dorso-ventrally flattened, kept folded back over the thorax in the resting position, moved forward through 180° for biting. Eyes either absent or present, with single or double lenses, with four lenses in one species. Dorsal ocelli absent. Antennae apparently 2-segmented, second segment invaginated into the first. Arista dendriform.

Thorax: Sternites fused in a broad plate. Dorsal surface membranous in its greater part. Two sclerotized ridges, the notopleural sutures, enclose a median membranous space, the mesonotum. Two pairs of thoracic spiracles, the anterior inside the notopleural sutures, the posterior near the haltere groove.

Wings: Absent, halteres present. A pair of movable ctenidia situated at the anterior part of the sides of the thorax in all except one subgenus.

Legs: Inserted on the dorsal surface, giving the insect a spider-like appearance. Empodium absent.

Abdomen: Consisting of seven segments and the anal segment. Seven pairs of abdominal spiracles. The segments are clearly marked on the dorsum of the male, but reduced in number by fusion or membranization on the venter of the male and on the abdomen of the female. A ctenidium is present on sternite 1+2, reduced in some genera and absent in one genus and in some species of other genera.

Genitalia: Not retracted inside the abdomen at rest, except in the genus *Cyclopodia*. Cerci (claspers) present, 1-segmented.

Eleven genera, about two hundred species, distributed throughout the world, mainly in the tropics and subtropics.

CLASSIFICATION

The Nycteribiidae are divided into two subfamilies, which are defined as follows:

NYCTERIBIINAE

Parasites of Microchiroptera. Head laterally compressed or rounded, never dorso-ventrally compressed. Eyes present or absent. Mesonotum parallel-sided or narrowing posteriorly. Tibiae laterally compressed, with several rows of long setae in the distal part of the ventral margin. Notopleural setae usually forming one or several diagonal rows, reduced or absent in only a few species or in one sex of a species. Mesopleural sutures originating half-way between the spiracles. Tergites 1 and 2 of abdomen usually separate, fused in the males of some species of *Penicillidia* and in both sexes of *Hershkovitzia*. Abdomen of female with two, three or four tergites before the anal segment, seven tergites and seven sternites in *Hershkovitzia*. Spines of thoracic and abdominal ctenidia narrow, pointed. Genitalia of *Nycteribia* type. Old and New Worlds. Six genera, of which four occur in the Palaearctic region and in Israel.

CYCLOPODIINAE

Parasites of Megachiroptera. Head laterally or dorso-ventrally compressed. Eyes always present. Mesonotum widening posteriorly. Mesopleural sutures originating far posteriorly, except in *Archinycteribia*. Tibiae cylindrical, with two or three bands of weaker integument and short setae in the middle. Notopleural setae usually reduced in number, rarely absent. Tergites 1 and 2 of abdomen fused in all species in both sexes. Segmentation of abdomen of female markedly reduced. Setae coarse, spines of ctenidia coarse and blunt. Genitalia of *Nycteribia*, *Eucampsipoda* and *Cyclopodia* type. Old World only. Five genera, of which only one occurs in the Palaearctic region and in Israel.

Diptera Pupipara

Key to the Palaearctic Genera and Subgenera of Nycteribiidae

1. Eyes absent   2
   – Eyes present   4

2. Tibiae short, two to three and a half times as long as wide. Row of setae at posterior margin of sternal plate of thorax complete or with a small gap in the middle. Dorsal genital plate of female with spines. Praegonites movably articulated with phallobase   3
   – Tibiae four and a half to five times as long as wide. Row of setae at posterior margin of sternal plate of thorax reduced to a few setae or a single seta at each side. Dorsal genital plate of female shield-shaped, with a few minute hairs. Ventral genital plate absent. Praegonites fused with the phallobase.
   Stylidia Westwood

3. Tibiae nearly semicircular, two to two and a half times as long as wide. Dorsal and ventral genital plates present or ventral plate absent.
   Nycteribia Latreille, subgenus **Nycteribia** s. str.
   – Tibiae slender, three and a half times as long as wide. Only a dorsal genital plate present.   **Nycteribia,** subgenus **Acrocholidia** Kolenati

4. Eyes single-faceted, unpigmented, projecting little from the surface   5
   – Eyes double-faceted, pigmented, markedly projecting from the surface. Tergal plate 2 of abdomen of female very large, either with two posterior processes with long setae or with straight or concave posterior margin.
   Basilia Miranda-Ribeiro, subgenus **Basilia** s. str.

5. Medium-sized or large species, 3–5 mm. Head rounded, helmet-shaped. Labella of labium as long as the theca. Mesonotum narrowing posteriorly. Haltere groove closed by cover. Fore coxae broad and short. Meso-metasternal sutures closed. Tibiae slender, with several rows of setae at the distal end. Spines of thoracic ctenidia slender, pointed. Postspiracular sclerite absent. Genitalia of *Nycteribia* type.   **Penicillidia** Kolenati
   – Smaller species, 3 mm. Head laterally compressed. Labella of proboscis twice as long as the theca. Mesonotum widening posteriorly. Haltere groove open. Fore coxae long, conical. Meso-metasternal sutures open. Tibiae cylindrical, with two bands of weaker integument and short setae in the middle. Postspiracular sclerite present. Genitalia of *Eucampsipoda* type.
   Eucampsipoda Kolenati

Subfamily NYCTERIBIINAE

DIAGNOSIS

Parasites of Microchiroptera. Head laterally compressed or rounded, never dorso-ventrally compressed. Eyes present or absent. Mesonotum parallel-sided or narrowing posteriorly. Tibiae laterally compressed, with several rows of long setae in the distal part of the ventral margin. Notopleural setae usually forming

124

one or several diagonal rows, reduced or absent in only a few species or in one sex of a species. Mesopleural sutures originating half-way between the spiracles. Tergites 1 and 2 of abdomen usually separate, fused in the males of some species of *Penicillidia* and in both sexes of *Hershkovitzia*. Abdomen of female with two, three or four tergites before the anal segment, seven tergites and seven sternites in *Hershkovitzia*. Spines of thoracic and abdominal ctenidia narrow, pointed. Genitalia of *Nycteribia* type. Old and New Worlds. Six genera, of which four are Palaearctic and occur in Israel.

Genus N Y C T E R I B I A Latreille, 1796
Précis des caractères génériques des insectes, p. 176

Type Species: *Nycteribia pedicularia* Latreille, 1805.

Small or medium-sized species with laterally compressed head. Eyes absent. Notopleural plates broad, with a row of six of fifteen notopleural setae. Tibiae laterally compressed, two to three and a half times as long as wide. Haltere groove open. Meso-metasternal sutures open. Abdomen of female with three or four tergites before the anal segment. Postspiracular sclerite narrow, with several setae. Dorsal genital plate with spines, ventral plate without spines, or only a dorsal genital plate present or both plates absent. Praegonites movable.

Key to the West Palaearctic Species of Nycteribia

1. Head membranous in the anterior part of the dorsal surface. Tibiae very broad, nearly semicircular, two to two and a half times as long as wide. Tergite 1 with short setae at the posterior margin (subgenus *Nycteribia*)   2
 - Head sclerotized to the anterior margin. Tibiae slender, three and a half times as long as wide. Tergite 1 with long setae at the posterior margin. Aedeagus of male broad and short, with rounded end. Dorsal genital plate of female concave posteriorly, with four to six widely spaced spines. Ventral genital plate absent (subgenus *Acrocholidia*).   **N. vexata** Westwood

2. Three tergites before the anal segment in the female. Dorsal genital plate triangular, ventral genital plate present. Anal segment of male short, conical. Sternite 5 of male with straight or slightly concave posterior margin   3
 - Four tergites before the anal segment in the female. Tergite 3 elliptical, with long setae. Dorsal genital plate of female elliptical, with a few short setae. Ventral genital plate absent. Sternite 5 of male strongly convex posteriorly, with fourteen to sixteen short spines. Anal segment very long, nearly parallel-sided. Aedeagus very slender apically, with bifid, recurved tip, and with scales on the dorsal membrane.   **N. schmidlii** Schiner

3. Length 3 mm. Tergite 2 of female long, convex posteriorly. Dorsal genital plate triangular, with truncate sides and eleven to fifteen short spines. Ventral plate

broad. Sternite 5 of male concave posteriorly, with fourteen to eighteen spines at the posterior margin. Aedeagus tapering, narrow apically and with a short ventral tooth near the end. **N. latreillii** Leach

– Length 2.5 mm. Tergite 2 of female short, slightly curved posteriorly. Pleurae bare. Dorsal genital plate triangular, with about twelve short spines. Ventral plate narrow. Posterior margin of sternite 5 of male nearly straight and with nine to twelve spines in the middle. Phallobase with a dorsal bulge. Aedeagus short, convex dorsally, with broad, rounded end and a ventral tooth at the distal third. **N. pedicularia** Latreille

– Length 2.0–2.5 mm. Tergite 2 of female convex or pointed posteriorly. Genital plates as in *N. pedicularia*, dorsal plate with eight to ten long setae at the posterior margin. Sternite 5 of male with seven or eight spines at the posterior margin. Phallobase conical. Aedeagus slender, with a ventral tooth at the distal fifth (continental Europe). **N. kolenatii** Theodor and Moscona

N. lindbergi from Afghanistan not included.

Subgenus **Nycteribia** s. str.

Listropoda Kolenati, 1857. *Wiener Ent. Mschr.*, 1 : 61.

Head partly membranous in the anterior dorsal part. Tibiae short and broad, laterally compressed, with three rows of long setae in the distal part of the ventral edge. Either a dorsal and ventral genital plate present or only a dorsal plate, or both absent.

The subgenus is divided into three groups:

1. *Pedicularia* group, with three tergites before the anal segment in the female and with two genital plates. Palaearctic and Oriental regions.

2. *Schmidlii* group, with four tergites before the anal segment in the female. Only a dorsal genital plate. Ethiopian, one species in the Mediterranean.

3. *Parilis* group, with four tergites before the anal segment in the female, a very short second tergite, sternites 7 and 8 fused. Genital plates absent. Cerci of male with a marked angle near the base. Oriental and Pacific regions.

Nycteribia pedicullaria Latreille, 1805
Figures 259, 270–274

Nycteribia pedicularia Latreille, 1805. *Histoire naturelle des Crustacées et des Insectes*, XIV, p. 403.
Nycteribia pedicularia Latreille. Theodor & Moscona (1954) *Parasitology*, 44 : 176.
Nycteribia pedicularia Latreille. Theodor (1954) in Lindner, 66a, Nycteribiidae, p. 19; (1967) *Cat. Rothsch. Coll. Nyct.*, p. 78.

Length 2.3–2.5 mm.
Head: Two to four setae at the anterior margin. Labella of proboscis as long as the theca.

Thorax: Wider than long. Length to width = 7 : 9. Meso-metasternal sutures forming an angle of 90°. Six to nine notopleural setae. Tibiae with three rows of long setae in the distal half of the ventral margin, two and a quarter times as long as wide.

Male Abdomen: Tergite 1 with short setae at the posterior margin. Tergites 2 and 3 with marginal rows of longer and shorter setae and short hairs in the middle of the surface. Tergites 4–6 with marginal rows of still longer setae and short spines between them, bare on the surface, except for a few hairs on tergite 4 in some specimens. Abdominal ctenidium with about fifty spines. Sternite 5 usually with nine to twelve spines at the posterior margin, rarely seven to eight or thirteen to fourteen. *Genitalia:* Cerci straight, with dark, blunt ends. Phallobase with a pronounced dorsal bulge. Aedeagus short, and broad, with a rounded end and a ventral tooth at the distal third.

Female Abdomen: Tergite 1 as in the male. Tergite 2 with a marginal row of five or six long setae in the middle and shorter setae and spines betwen the long setae and laterally. Tergite 2 broadly rounded and shorter in the middle than the

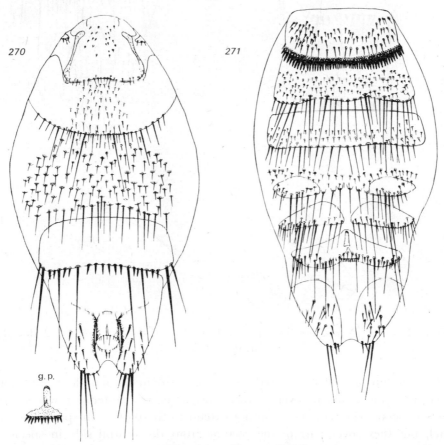

Figs. 270–271: *Nycteribia pedicularia* Latreille. Female abdomen.
270. dorsal, with genital plate (g.p.); 271. ventral

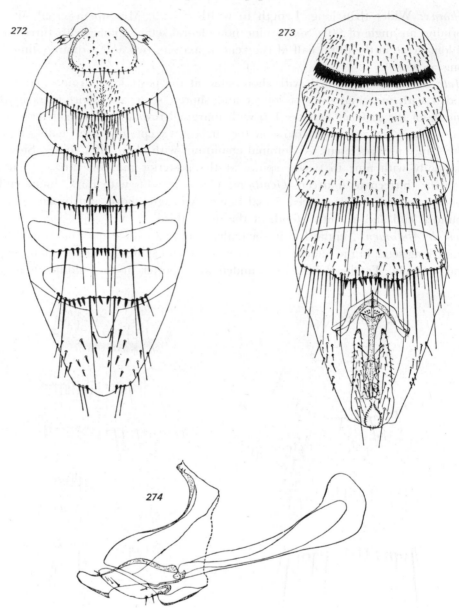

Figs. 272–274: *Nycteribia pedicularia* Latreille. Male abdomen. 272. dorsal; 273. ventral; 274. genitalia

width of tergite 1. Surface covered with short hairs in the middle. Connexivum between tergites 2 and 6 covered with short hairs, the posterior row consisting of longer setae. The short hairs do not extend to the pleurae in specimens from Israel, but they cover part of the pleurae from the ventral side in specimens from Europe. Tergite 6 wide, bare, with six to eight long setae at the posterior margin and about fifteen spines in groups between the setae. Sternite 1+2 as in the male. Sternite 5 with elliptical lateral sclerites and long setae posteriorly and four or five setae between the sclerites. Sternite 6 with similar sclerites, but they are larger and reach to near the midline. Sternite 7 triangular, with a double row of setae. Dorsal genital plate triangular, with a row of about twelve short spines posteriorly. Ventral plate very narrow.

Hosts: Species of *Myotis, Miniopterus, Rhinolophus* and other genera.

Distribution: Continental Europe; S.W. Asia; N. Africa.

Israel: Galilee, Coastal Plain, about one hundred specimens, on *Myotis myotis* Borkhausen, *M. nattereri* Kuhl, *M. mystacinus* Kuhl, *Miniopterus schreibersi* Kuhl; May to September.

Nycteribia latreillii (Leach, 1817)
Figures 216–221, 229–230, 234–235, 237, 241–246, 252, 258, 265, 275

Phthiridium latreillii Leach, 1817. *Zoological Miscellany*, III, p. 54.
Listropodia latreillii Leach. Kolenati (1863) *Horae Soc. Ent. Ross.*, 2 : 55.
Nycteribia biscutata Gil Collado, 1934. *Eos*, 9 : 29 (abnormal specimen).
Nycteribia latreillii africana Karaman, 1939. *Annls Musei Serb. Merid.*, 1 : 31.
Nycteribia latreillii Leach. Theodor & Moscona (1954) *Parasitology*, 44 : 179.
Nycteribia latreillii Leach. Theodor (1954) in Lindner, 66a, Nycteribiidae, p. 17;
(1967) *Cat. Rothsch. Coll. Nyct.*, p. 69.

Length 3 mm. Colour brown.

Head and Thorax: Similar to that of *N. pedicularia*. Nine to twelve notopleural setae.

Male Abdomen: Tergite 4 with a large group of short setae in the middle of the surface and a few setae on tergites 5 and 6. Sternite 5 with an irregular row of fourteen to eighteen spines at the posterior margin, which is concave in the middle. The lateral spines are very long. All setae longer and stronger than in *N. pedicularia*. *Genitalia*: Phallobase conical, without dorsal bulge. Aedeagus tapering to a rounded point, with a ventral tooth near the end.

Female Abdomen: Tergite 2 as long as or longer than the width of tergite 1, strongly convex posteriorly, with six to eight long setae in the middle of the posterior margin, short spines between them and shorter setae laterally. A large group of short hairs in the middle of the surface. The short setae on the pleurae reach to tergite 6. Tergite 6 as in *N. pedicularia*. Dorsal genital plate triangular, with truncate sides and twelve to fifteen strong spines posteriorly. Ventral plate broad.

Hosts: Species of *Myotis, Miniopterus, Rhinolophus* and other genera.

Distribution: Continental Europe; S.W. Asia to Afghanistan and Kirghizia; N. Africa.

Israel: N. Galilee, about twenty specimens, on *Myotis myotis*; August to October.

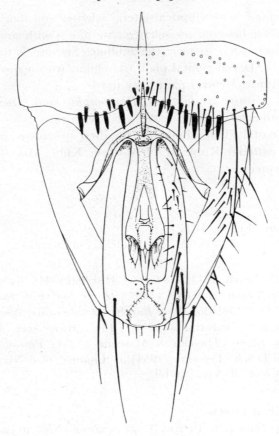

Fig. 275: *Nycteribia latreillii* (Leach). Male, sternite 5 and genital area

Nycteribia schmidlii Schiner, 1853
Figures 276–282

Nycteribia schmidlii Schiner, 1853. *Verh. Zool.-Bot. Ges. Wien,* 3 : 151.
Nycteribia blasii Kolenati, 1856 (nec 1863). *Die Parasiten der Chiropteren,* Brünn, p. 41.
Listropodia schmidlii Schiner. Kolenati (1863) *Horae Soc. Ent. Ross.,* 2 : 52.
Nycteribia schmidlii Schiner. Theodor & Moscona (1954) *Parasitology,* 44 : 185.
Nycteribia schmidlii Schiner. Theodor (1954) in Lindner, 66a, Nycteribiidae, p. 20; (1967) *Cat. Rothsch. Coll. Nyct.,* p. 90.

Length 2.0–2.25 mm. Colour brown.
Head: Setae at the anterior dorsal margin of the head standing close together.
Thorax: As wide as long. Mesonotum narrow, six to nine notopleural setae.
Male Abdomen: Slender and narrow, held curved downwards. Tergites 1–4 with short setae on the surface. Tergites 2 and 3 with moderately long setae in

130

the marginal rows. Tergite 4 more strongly convex posteriorly and the setae on the surface longer. Marginal row with longer setae, two setae in the middle of the row very long. Tergite 5 similar, but surface bare and two to four setae in the middle of the marginal row very long. Tergite 6 very narrow, strip-like, bare, usually covered by tergite 5. Its marginal row consists of moderately long, thin setae. Anal segment very long and slender, as long as the four preceding tergites together, nearly parallel-sided. Abdominal ctenidium with forty to fifty spines. Sternite 5 strongly convex posteriorly, with a double row of fourteen to sixteen short, barrel-shaped spines at the posterior margin. *Genitalia*: Cerci curved, with dark ends. Phallobase conical. Aedeagus long, with a narrow, recurved, bifid tip. Dorsal membrane with scale-like teeth in the basal half and ending in two pointed processes posteriorly.

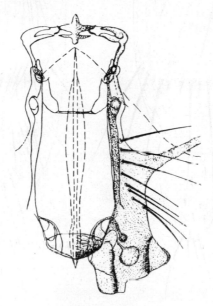

Fig. 276: *N. schmidlii* Schiner. Thoracic pattern

Female Abdomen: Tergite 1 with short setae at the posterior margin. Tergite 2 shorter in the middle than the width of tergite 1, with nearly straight posterior margin which bears a row of short, thin setae and with short hairs on the surface. Tergite 3 elliptical, with six very long setae and eight to ten spines at the posterior margin and short hairs on the surface. Tergite 6 very wide, bare, with eight to ten long setae and ten to twelve long spines at the posterior margin. Connexivum between tergites 3 and 6 covered with short hairs, pleurae bare. A group of spines laterally and posteriorly to tergite 6. Anal segment divided into two lobes. Sternites 5 and 6 divided into elliptical sclerites with two rows of setae, which continue between the sclerites. Sternite 7 also divided into lateral sclerites which are placed close together. Dorsal genital plate elliptical, with four to eight widely spaced, short setae posteriorly. Ventral plate absent.

131

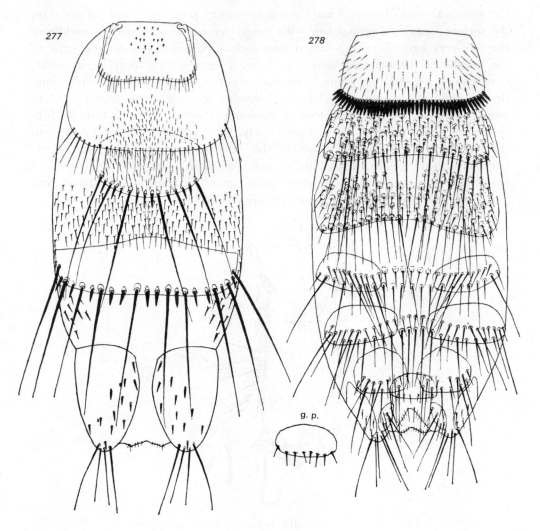

Figs. 277–278: *Nycteribia schmidlii* Schiner. Female abdomen. 277. dorsal, with genital plate (g.p.); 278. ventral

Hosts: Mainly *Miniopterus schreibersi*, more rarely species of *Myotis* and *Rhinolophus*.
Distribution: C. and S. Europe; S. W. Asia to Afghanistan; N. Africa.
Israel: Galilee to southern Coastal Plain; common on *Miniopterus schreibersi*, rare on *Myotis myotis*; June to September, mainly in August and September.

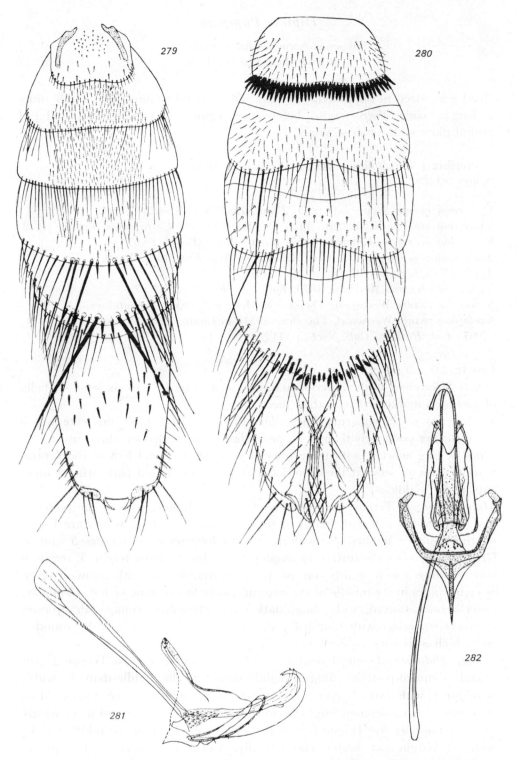

Figs. 279–282: *Nycteribia schmidlii* Schiner. Male. 279. abdomen, dorsal;
280. abdomen, ventral; 281. genitalia, lateral; 282. genitalia, ventral

Subgenus **Acrocholidia** Kolenati, 1857
Wiener Ent. Mschr., 1 : 62.

Head sclerotized to the anterior margin. Tibiae slender, three and a half times as long as wide. Three tergites before anal segment of female. Only a dorsal genital plate with spines or setae.

Nycteribia (Acrocholidia) vexata Westwood, 1835
Figures 283–286

Nycteribia vexata Westwood, 1835. *Trans. Zool. Soc. Lond.*, 1 : 291.
Nycteribia montaguei Kolenati, 1856. *Die Parasiten der Chiropteren*, Brünn, p. 38.
Nycteribia bechsteini Kolenati, 1857. *Wiener Ent. Mschr.*, 1 : 62.
Acrocholidia montagui Kolenati, 1863. *Horae Soc. Ent. Ross.*, 2 : 60.
Acrocholidia bechsteini Kolenati, 1863. *Ibid.*, p. 63.
Nycteribia ercolanii Rondani, 1879. *Boll. Soc. Ent. Ital.*, 11 : 7.
Nycteribia vexata Westwood. Theodor & Moscona (1954) *Parasitology*, 44 : 190.
Nycteribia vexata Westwood. Theodor (1954) in Lindner, 66a, Nycteribiidae, p. 21; (1967) *Cat. Rothsch. Coll. Nyct.*, p. 112.

Length 2.0–2.5 mm.
Head: Sclerotized to the anterior margin which bears four to six setae. Labella of proboscis markedly longer than theca.
Thorax: As in the subgenus *Nycteribia*. Eight to ten notopleural seate which begin further posteriorly than in *N. pedicularia*. Tibiae slender, three and a half times as long as wide, with three rows of setae in the distal half of the ventral margin. Row of setae at the posterior margin of the sternal plate with a small gap in the middle.
Male Abdomen: Tergite 1 with a row of moderately long setae at the posterior margin. Tergites 2–6 with marginal setae of uniform length, which are longer on the posterior tergites, particularly two to four setae on tergites 5 and 6. Groups of hairs on the surface of tergites 2–4, a few hairs on tergite 5, tergite 6 bare. Sternite 5 with slightly concave posterior margin and with a row of seven or eight spines in the middle of the margin, rarely five or nine or ten. *Genitalia*: Cerci slender, curved, with blunt, dark ends. Phallobase conical. Praegonites narrowly triangular with rounded end. Aedeagus short, with broadly rounded end which is as wide as the basal part.
Female Abdomen: Tergite 1 with a row of long setae posteriorly. Tergite 2 with broadly rounded posterior margin, slightly shorter in the middle than the width of tergite 1, with a small group of short hairs in the middle of the surface. Marginal row with moderately long setae in the middle and shorter and more widely spaced setae laterally. Tergite 6 with five or six long setae in the middle of the posterior margin and shorter setae laterally. A few short hairs on the surface. Connexivum between tergites 2 and 6 with hairs which are less numerous than in *N. pedicularia*. Anal segment with long posterior processes, with setae at the

end. Abdominal ctenidium with forty-five to fifty spines. Sternites 5 and 6 divided into lateral sclerites with long setae in the marginal rows. Sternite 7 undivided, rectangular. Dorsal genital plate irregularly rounded, with sclerotized posterior margin and four to six short spines. Ventral genital plate absent.

Hosts: Mainly species of *Myotis,* also species of *Rhinolophus* and other genera.

Distribution: Continental Europe, S.W. Asia; N. Africa.

Israel: N. Galilee, about twenty specimens; on *Myotis myotis*; June to October.

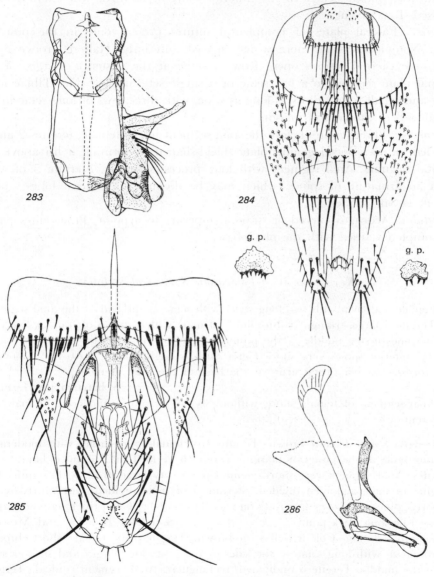

Figs. 283–286: *Nycteribia vexata* Westwood: 283. thoracic pattern;
284. female abdomen, dorsal, and genital plates (g.p.); 285. male,
sternite 5 and genital area; 286. male genitalia

Diptera Pupipara

Genus S T Y L I D I A Westwood, 1840
Introduction to the Modern Classification of Insects, II, p. 154.

Phthiridium Hermann, 1804. *Mémoire Aptèrologique*, p. 120 (synonym of *Nycteribia*).
Celeripes Montagu, 1808. *Trans. Linn. Soc. Lond.*, 9 : 166 (nomen nudum).

Type Species: *Stylidia biarticulata* Hermann, 1904.
Head: Sclerotized nearly to the anterior margin, more or less laterally compressed. Eyes absent.
Thorax: Lateral plates of notopleural sutures present only in the posterior half. Notopleural setae more widely spaced anteriorly. Haltere groove open. Meso-metasternal sutures open. Row of setae at the posterior margin of the sternal plate reduced to a few setae or a single seta at each side. Tibiae long, four and a half to five times as long as wide, with three rows of long setae in the distal half of the ventral margin.
Abdomen: Three tergites before the anal segment in the female, tergites 2 and 3 divided or not. Dorsal genital plate shield-shaped or forming a lip above the genital opening. Anal segment with long processes or not. Sternite 5 of male with an armature of spines, which may be divided into lateral lobes; spines absent in one species.
Genitalia: Aedeagus curved or straight, tapering to a point. Praegonites partly or completely fused with the phallobase.

Key to the West Palaearctic Species of Stylidia

1. Female abdomen with two long styles with setae at the end of the anal segment. Tergite 6 large, triangular. Sternite 5 of male with a group of four or five rows of spines in the middle of the posterior margin, the posterior spines very long, the anterior spines very short. Cerci thick, straight, blunt. Two bare, movable processes on sclerotized arms at the base of the anal segment.

S. biarticulata Hermann

 – Anal segment of female short, without styles. Sternite 5 and cerci of male different 2

2. Length 3 mm. Colour brown. Tergite 2 of female very large, with moderately long setae at the posterior margin. Tergite 6 divided into rounded lateral sclerites. Anal segment very short. Genital plate very large, 0.35 × 0.27 mm. Tergites of male abdomen divided. Sternite 5 of male divided, with thirty-five to forty spines in four or five rows on each lateral lobe. Cerci thick, curved, tapering to a long, dark point. **S. biloba** Theodor and Moscona
   Length 2.5 mm. Colour yellowish brown. Tergite 2 of female short, broadly rounded, with long setae at the sides of the posterior margin and shorter setae in the middle. Tergite 6 undivided, rectangular. Anal segment conical. Tergites of male abdomen undivided. Sternite 5 undivided, with straight posterior margin with a double row of about twenty-five spines. Cerci slender, slightly curved. **S. integra** Theodor and Moscona

136

Stylidia biarticulata (Hermann, 1864)

Figures 214–215, 287–290

Phthiridium biarticulatum Hermann, 1804. *Mémoire Aptèrologique,* p. 124.
Celeripes vespertilionis Montagu, 1808. *Trans. Linn. Soc. Lond.,* 9 : 166.
Phthiridium hermanni Leach, 1817. *Zoological Miscellany,* III, p. 55.
Nycteribia hermanni Leach. Kolenati (1856) *Die Parasiten der Chiropteren,* Brünn, p. 37.
Stylidia hermanni Leach. Kolenati (1863) *Horae Soc. Ent. Ross.,* 2 : 66.
Stylidia biarticulata Hermann. Westwood (1840) *Introduction to the Modern Classification of Insects,* II, p. 154.
Nycteribia (Stylidia) biarticulata Hermann. Theodor & Moscona (1954) *Parasitology,* 44 : 193; Theodor (1954) in Lindner, 66a, Nycteribiidae, p. 22.
Stylidia biarticulata Hermann. Theodor (1967) *Cat. Rothsch. Coll. Nyct.,* p. 122 (full synonymy).

Length 2.5–3.0 mm. Colour brown.

Thorax: As long as wide. Meso-metasternal sutures forming an angle of 75°. Ten to thirteen notopleural setae, which are more widely spaced anteriorly. Posterior margin of sternal plate with four setae at each side, one of them very long.

Male Abdomen: Tergite 1 with a marginal row of short setae with a gap in the middle. Tergites 2–6 with marginal rows of moderately long setae, two to four long setae in the marginal rows of tergites 5 and 6. Short hairs on the surface of tergites 2–5, tergite 6 bare. Anal segment wide and rounded posteriorly, with a deep, angular incision of the anterior dorsal margin. Abdominal ctenidium with about forty-five spines. Sternite 5 with convex posterior margin with a group of forty to fifty spines in four or five rows, the posterior spines much longer than the anterior spines. Two rounded processes with reticulate surface at the base of the anal segment, which are situated on a sclerotized arm which can be moved outwards. Two rounded processes with numerous setae posterior to the processes. *Genitalia*: Cerci thick, straight, with blunt, dark ends. Phallobase concave dorsally. Aedeagus curved, tapering to a rounded, knob-like end. Praegonites broad, with a curved apical process and with four or six setae at its base, fused with the phallobase.

Female Abdomen: Tergite 1 concave posteriorly, each lobe with a marginal row of short setae at each side, leaving the median concavity bare. Tergite 2 large, rounded, incompletely divided in the middle, much longer than the width of tergite 1. The marginal row consists of long setae laterally, and shorter setae and spines in the middle. Connexivum between tergites 2 and 6 bare, except for a group of eight to twelve short setae in the middle. Tergite 6 triangular, with a few long setae at the apex and a few short hairs on the surface. Anal segment membranous dorsally, with two sclerites ventrally which are prolonged into long styles with four setae at the end which form a cross. A few short hairs at the stem of the styles. Sternites 5 and 6 divided into narrow lateral sclerites which nearly reach the midline and bear two or three long setae at the lateral

137

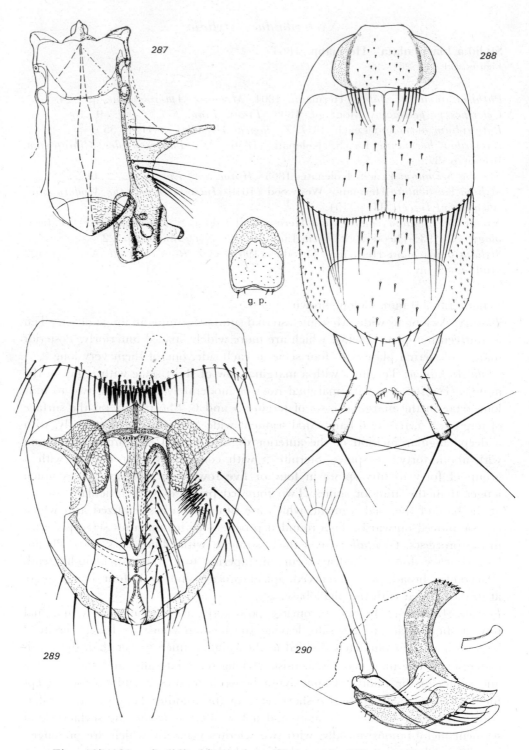

Figs. 287–290: *Stylidia biarticulata* (Hermann). 287. thoracic pattern;
288. female abdomen, dorsal, and genital plate (g.p.); 289. male, genital
area and sternite 5; 290. male genitalia

corners. Sternite 7 longer, also divided into lateral sclerites which are roughly triangular and bear four or five long setae posteriorly. Genital plate shield-shaped, 0.15 × 0.20 mm, with more strongly sclerotized posterior margin, which projects from the surface; a few minute hairs at the base.

Hosts: Mainly species of *Rhinolophus*, rarely other genera.

Distribution: Europe, including British Isles; N. Africa; S.W. Asia to Afghanistan and Kirghizia.

Israel: Galilee to southern Coastal Plain, Jerusalem; on *Rhinolophus ferrum-equinum* Schreber, *R. hipposideros minimus* Heuglin, *R. euryale judaicus* And. and Matsch.; *R. blasii* Peters, *Myotis myotis, Miniopterus schreibersi* (rare); throughout the year.

Stylidia biloba (Theodor and Moscona, 1954)
Figures 291–296

Nycteribia (Stylidia) biloba Theodor & Moscona, 1954. *Parasitology*, 44 : 195; Theodor (1954) in Lindner, 66a, Nycteribiidae, p. 23; *Cat. Rothsch. Coll. Nyct.*, p. 127.

Length 2.75 mm. Colour brown.
Head and Thorax: As in *S. biarticulata*, except for minor differences.

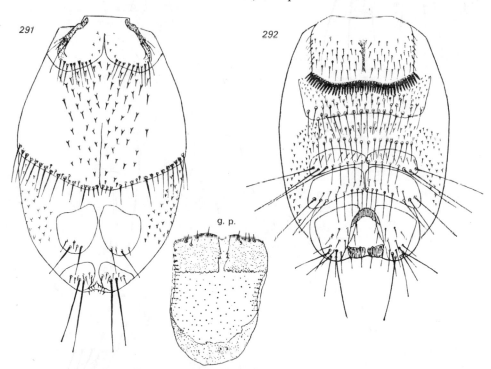

Figs. 291–292: *Stylidia biloba* (Theodor and Moscona). Female abdomen.
291. dorsal, with genital plate (g.p.); 292. ventral

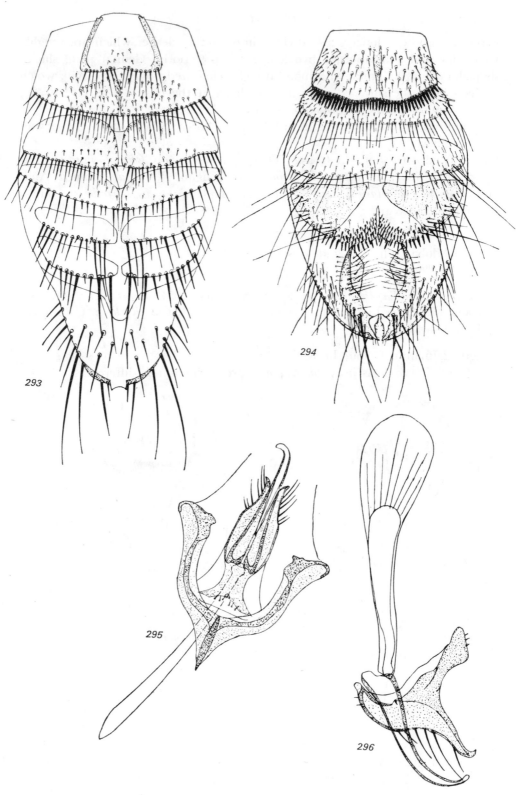

Figs. 293–296: *Stylidia biloba* (Theodor and Moscona). Male. 293. abdomen, dorsal;
294. abdomen, ventral; 295. genitalia, ventral; 296. genitalia, side view

Male Abdomen: Tergites 2–6 more or less distinctly divided in the middle. Tergites 2–4 with marginal rows of short setae and short hairs on the surface. Tergites 5 and 6 with longer setae in the middle of the marginal rows and bare on the surface. Anal segment broadly rounded, with a deep, rounded indentation of the anterior dorsal margin. Sternite 5 with two broad lobes posteriorly, each with four or five rows of short spines, the spines of the posterior row longer, but not as long as in *S. biarticulata*. *Genitalia*: Cerci thick, curved, tapering to a long, dark point. Praegonites with a curved apical process which is much longer than in *S. biarticulata*, with five or six setae ventrally, fused with the phallobase. Aedeagus long, curved, tapering to a rounded point.

Female Abdomen: Tergite 1 divided into two lobes, each with a marginal row of about eight setae. Tergite 2 very large, incompletely divided in the middle, with a marginal row of moderately long setae laterally, shorter setae and spines in the middle. A large group of short hairs in the middle of the surface. Tergite 6 divided into two closely placed, rounded sclerites, each of which bears one or two longer setae and some spines posteriorly. Connexivum between tergites 2 and 6 bare. Anal segment very short, broadly conical, with two setae at the end of each lateral sclerite. Abdominal ctenidium with about fifty-five spines. Sternites 5 and 6 divided into lateral sclerites which reach to the midline, with marginal rows of short setae and two or three vertical long setae laterally. Sternite 7 divided into two triangular plates with setae posteriorly. Genital plate very large, shield-shaped, 0.35 × 0.27 mm, with more heavily sclerotized anterior and posterior margins, and a few minute hairs at the base.

Hosts: Species of *Rhinolophus*, *Miniopterus* and *Myotis*.

Distribution: The subspecies *S. biloba orientalis* has been described from Afghanistan by Hurka and Povolny (1968); however this is apparently not a subspecies of *S. biloba*, but a distinct species.

Israel: Galilee to southern Coastal Plain, Jerusalem; September to February. Herzliyya (8), holotype male from *Rhinolophus hipposideros minimus*; 5 February 1950. Three male, three female paratypes from *R. ferrumequinum, Miniopterus schreibersi* and *Myotis* sp., in the Department of Parasitology, Hebrew University of Jerusalem. One male, one female paratype in the British Museum (Nat. Hist.).

Stylidia integra (Theodor and Moscona, 1954)
Figures 297–303

Nycteribia blainvillii Leach. Speiser (1901) *Arch. Naturgesch.*, 67 : 34 (misdetermination).
Nycteribia blainvillii Leach. Karaman (1939) *Annls Musei Serb. Merid.*, 1 : 31 (misdetermination).
Nycteribia (Stylidia) integra Theodor & Moscona, 1954. *Parasitology*, 44 : 198 (nom. nov.); Theodor (1954), in Lindner, 66a, Nycteribiidae, p. 25.
Stylidia integra Theodor & Moscona. Theodor (1967) *Cat. Rothsch. Coll. Nyct.*, p. 153.

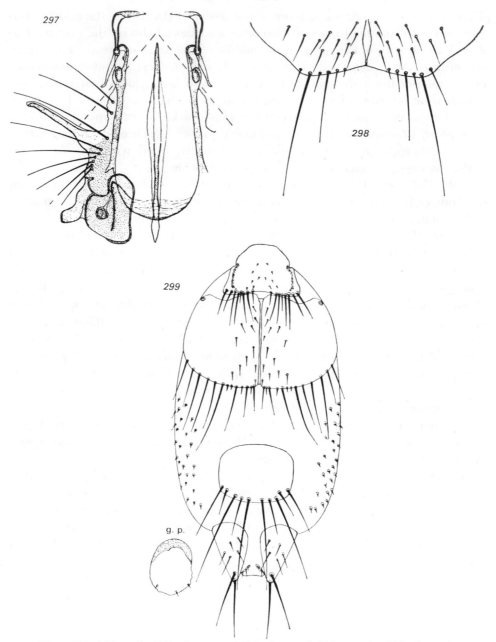

Figs. 297–299: *Stylidia integra* (Theodor and Moscona). 297. thoracic pattern; 298. posterior margin of sternal plate; 299. female abdomen, dorsal, with genital plate (g.p.)

Length 2.5 mm. Colour yellowish brown.

Head and Thorax: As in *S. biarticulata*.

Male Abdomen: Tergites undivided, with marginal rows of moderately long setae and with two long setae in the middle of the marginal rows of tergites 5 and 6. About ten short hairs on the surface of tergite 3 and a few on tergites 4 and 5. Sternite 5 with straight or slightly convex posterior margin with a double row of about twenty-five spines. *Genitalia*: Cerci slender. Aedeagus tapering, slightly curved. Praegonites with a pointed apical process and three to five setae in the middle of the ventral margin, fused with the phallobase.

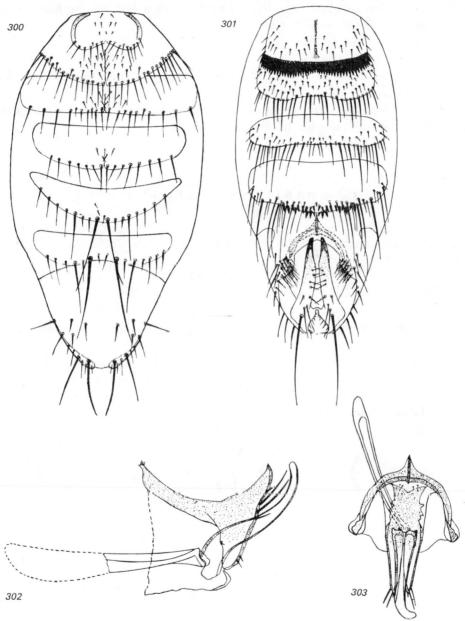

Figs. 300–303: *Stylidia integra* (Theodor and Moscona). Male. 300. abdomen, dorsal; 301. abdomen, ventral; 302. genitalia, side view; 303. genitalia, ventral

Female Abdomen: Tergites 1, 2 and 6 with marginal rows of long setae. Tergite 2 broadly rounded, with a marginal row of about eighteen setae, which are shorter in the middle. Tergite 6 rounded-rectangular, bare on the surface and with six long and some shorter setae at the posterior margin. Anal segment rather long, conical. Abdominal ctenidium with fifty to fifty-five spines. Dorsal genital plate shield-shaped, rounded anteriorly, with a few minute hairs at the base, much smaller than in *S. biloba*.

Hosts: Species of *Rhinolophus* and *Hipposideros*.

Distribution: Arabia, Egypt.

Israel: Jerusalem, one male; Cave of Twins (10), one male, from *Rhinolophus blasii*; February.

About a hundred and fifty specimens from S. Arabia and Egypt, mainly from *Rhinolophus acrotis* Heuglin, have been examined.

Genus B A S I L I A Miranda-Ribeiro, 1903
Archos Mus. Nac. Rio de J., 12 : 175.

Pseudelytromyia Miranda-Ribeiro, 1907. *Archos Mus. Nac. Rio de J.*, 14 : 231.
Guimaraesia Schuurmans-Stekhoven, 1951. *Acta Zool. Lilloana*, 12 : 101.

Type Species: *Basilia ferruginea* Miranda-Ribeiro, 1903.

Head: Laterally compressed, sclerotized to the anterior margin or with a small membranous area anteriorly. Eyes consisting of two facets on a common pigmented base (Figs. 304–310), rarely reduced to a single facet, absent in the subgenus *Tripselia*. Labella of proboscis shorter than the theca in most species.

Thorax: Usually broader than long. Tibiae more or less slender, a complete row of setae at the posterior margin of the sternal plate. Meso-metasternal sutures open. Haltere groove open.

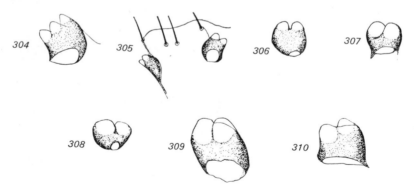

Figs. 304–310: Eyes of species of *Basilia*. 304–305. *B. daganiae* Theodor and Moscona; 306–308. *B. nana* Theodor and Moscona; 309–310. *B. nattereri* Kolenati

Abdomen: Postspiracular sclerite with several setae. Abdomen of female with two, three or four tergal plates before the anal segment. The tergal plates may be divided or not. Tergite 1 with two groups of long setae posteriorly in some species. Tergal plate 2 with long posterior processes and long setae in these species. Tergite 1 with a row of setae and tergal plate 2 with straight or rounded posterior margin in other species. Abdominal ctenidium very wide. Sternite 5 of male with spines at the posterior margin in most species. Genital plate reduced to a few setae on an area of microtrichia in most species.

Old and New Worlds.

About seventy-five species. Two species in Israel.

Key to the West Palaearctic Species of Basilia

1. Tibiae long, four and a half times as long as wide. Tergite 1 of female with two processes posteriorly with three to five very long, closely standing setae. Tergal plate 2 large, heart-shaped, with posterior processes which bear three to five long setae. Sternite 5 of male with fifteen to eighteen spines in two rows at the posterior margin. Aedeagus curved, tapering to a point.
   B. daganiae Theodor and Moscona

 -- Tibiae shorter, three to three and a half times as long as wide. Tergite 1 of female rounded posteriorly, with a row of long setae. Tergal plate 2 rectangular or trapezoidal. Aedeagus of male different   2

2. Tergite 1 of female with a row of long setae at the posterior margin which is broadly interrupted in the middle. Tergal plate 2 trapezoidal, narrower posteriorly, without sclerotized posterior corners. Anal segment with very long, slender processes. Sternite 5 of male with about thirty spines in two rows at the posterior margin. Aedeagus short and broad, with rounded end and a short, ventral apical tooth (C. and S. Europe).   **B. italica** Theodor
 -- Tergite 1 of female with a continuous row of setae at the posterior margin. Tergal plate 2 rectangular, with sclerotized posterior corners. Anal processes short, rounded. Sternite 5 of male with a row of six to eight spines at the posterior margin. Aedeagus different   3

3. Length 3 mm. Fifteen notopleural setae. Tergal plate 2 of female transverse-rectangular, covered with hairs in its greater part. Sternite 6 divided into two lateral, triangular sclerites. Aedeagus of male with long, membranous end which is divided into several processes. Dorsal membrane of aedeagus with scales. Praegonites with straight ventral margin (C. and S. Europe).
   B. nattereri Kolenati
 -- Length 1.8–2.2 mm. Nine to twelve notopleural setae. Tergal plate 2 of female square, with a diamond-shaped field of short hairs on the surface. Sternite 6 divided into three sclerites. Aedeagus of male with rounded, recurved point, serrated at the end. Dorsal membrane of aedeagus without scales. Praegonites with a large tooth at the ventral margin.   **B. nana** Theodor and Moscona
   B. afghanica Theodor not included.

145

Basilia daganiae Theodor & Moscona, 1954
Figures 304–305, 311–317

Basilia bathybothyra daganiae Theodor & Moscona, 1954. *Parasitology*, 44 : 207.
Basilia bathybothyra daganiae Theodor & Moscona. Theodor (1954) in Lindner, 66a, Nycteribiidae, p. 29.
Basilia daganiae Theodor & Moscona. Theodor (1967) *Cat. Rothsch. Coll. Nyct.*, p. 228.

Length 2.5 mm. Colour brown.
Head: Labella of proboscis one-third of the length of the theca.

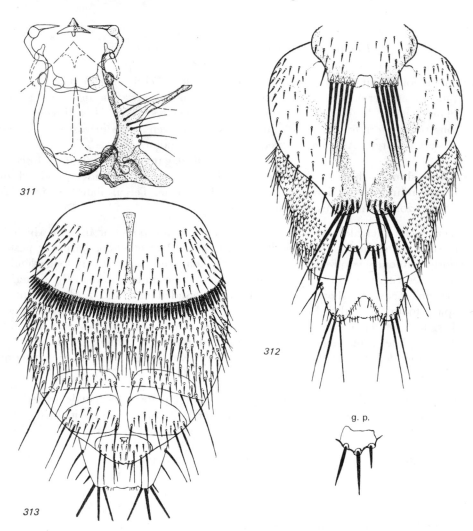

Figs. 311–313: *Basilia daganiae* Theodor and Moscona. 311. thoracic
pattern; 312. female abdomen, dorsal, with genital plate (g.p.);
313. female abdomen, ventral

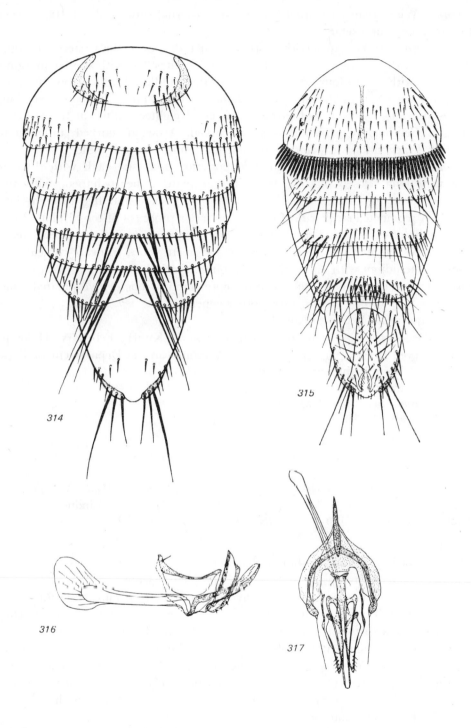

Figs. 314–317: *Basilia daganiae* Theodor and Moscona. Male. 314. abdomen, dorsal; 315. abdomen, ventral; 316. genitalia, side view; 317. genitalia, ventral

147

Thorax: Wider than long, angle of meso-metasternal sutures about 100°; seven to nine notopleural setae.

Male Abdomen: Tergite 1 with short setae at the sides of the posterior margin. Tergite 2 with a marginal row of short setae. Tergites 3–6 with similar marginal rows, but with two to four very long setae in the middle. Marginal row of tergite 6 with a wide gap in the middle. Abdominal ctenidium with about sixty long spines. Sternite 5 with a double row of fifteen to eighteen spines at the posterior margin. *Genitalia:* Cerci short, slightly curved. Aedeagus curved, tapering to a point. Praegonites with short triangular end and a ventral angle.

Female Abdomen: Tergite 1 with two posterior processes, each with three to five very long, closely placed setae. Tergal plate 2 heart-shaped, with two rounded posterior processes, each with three long setae and five to eight long spines. A pigmented stripe runs from the posterior processes of tergal plate 2 into the middle of each half. Tergal plate 3 small, with two rounded processes, each with a long seta and two short spines. Anal segment short, truncate-conical. Abdominal ctenidium with about sixty-five spines. Genital plate small, with two or three short setae. Anal sclerite absent or represented by one or two isolated setae.

Hosts: Species of *Pipistrellus*, rarely other genera.

Distribution: E. Mediterranean; Kenya.

Israel: Deganya 'A' (7); from *Pipistrellus kuhlii* (Kuhl); February. Holotype female, about twenty paratypes (coll. Y. Palmoni), in the Department of Parasitology, Hebrew University of Jerusalem.

Basilia nana Theodor and Moscona, 1954
Figures 306–308, 318–323

Basilia nana Theodor & Moscona, 1954. *Parasitology*, 44 : 204.
Penicillidia nattereri Kolenati, 1857. Speiser (1901) *Arch. Naturgesch.*, 67 : 40 (pro parte).
Listropodia nattereri Kolenati. Karaman (1948) *Rad. Jugosl. Akad.*, 273 : 117.
Basilia nana Theodor & Moscona, 1954. Theodor (1954) in Lindner, 66a, Nycteribiidae, p. 30; (1967) *Cat. Rothsch. Coll. Nyct.*, p. 209.

Length 1.8–2.2 mm.

Head: Labella of proboscis shorter than theca.

Thorax: Wider than long, nine to twelve notopleural setae.

Male Abdomen: Tergite 1 with concave posterior margin and a marginal row of short setae. Tergites 2 and 3 with marginal rows of moderately long setae. Tergites 4–6 with two to four long setae in the middle of the marginal rows. Abdominal ctenidium with fifty-five spines. Sternite 5 with a row of six to eight spines at the posterior margin. *Genitalia:* Cerci curved, with dark ends and a subapical tooth. Aedeagus with broadly rounded end and a recurved point. The rounded end is serrated. Praegonites narrowly triangular, with a strong tooth at the ventral margin.

Female Abdomen: Tergite 1 with a row of moderately long setae at the posterior margin which are shorter in the middle. Tergal plate 2 nearly square, with a

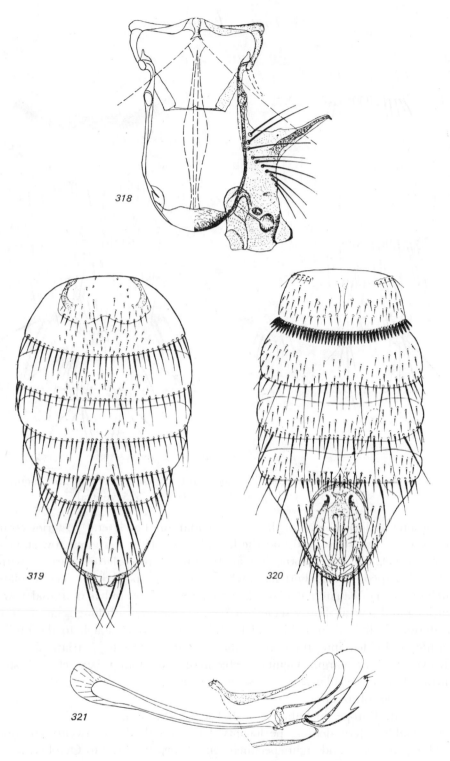

Figs. 318–321: *Basilia nana* Theodor and Moscona. 318. thoracic pattern;
319. male abdomen, dorsal; 320. male abdomen, ventral; 321. male genitalia

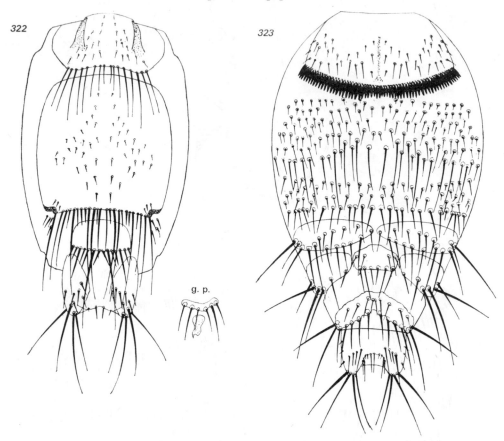

Figs. 322–323: *Basilia nana* Theodor and Moscona. Female abdomen.
322. dorsal, with genital plate (g.p.); 323. ventral

row of long and short setae at the posterior margin. The lateral posterior corners
are more strongly sclerotized and the lateral setae of the marginal row stand on
these corners. No setae at the lateral margins of the sclerite. Surface with a
diamond-shaped field of short hairs which are longer posteriorly. Tergal plate 3
much narrower than tergal plate 2, with bare surface and a marginal row of
four long and several shorter setae. Anal segment short, with long setae posteriorly. Sternite 5 divided into elliptical lateral sclerites which reach to the midline.
Sternite 6 divided into three triangular sclerites. Sternite 7 triangular, with
concave posterior margin. Genital plate small, with four to six setae. A small,
ventral plate of irregular shape in some specimens.

Hosts: Species of *Myotis*.

Distribution: Europe, including British Isles; S.W. Asia.

Israel: Galilee, Jerusalem (type locality); one hundred and twenty specimens
on *Myotis nattereri* and eight specimens on *M. myotis;* May to October, mainly
in July and August. Holotype male and numerous paratypes in the Department
of Parasitology, Hebrew University of Jerusalem.

150

Genus P E N I C I L L I D I A Kolenati, 1863
Horae Soc. Ent. Ross., 2 : 69.

Megistopoda Kolenati, 1857 (nec Macquart, 1852). *Wiener Ent. Mschr.*, 1 : 62.

Type Species: *Penicillidia dufourii* Westwood, 1835 (*Nycteribia*).*
Medium-sized or large species, 2.5–5.0 mm, generally very setose.
Head: Rounded posteriorly, with single, small, unpigmented eyes. Several rows of setae between the eyes on the vertex. Palps broad, with several rows of setae at the margin and ventrally and a long terminal seta.
Thorax: Wider than long. Mesonotum wide, narrower posteriorly. Notopleural setae present or absent. Haltere groove covered. Meso-metasternal sutures closed. Thoracic ctenidium with narrow, pointed spines, absent in the subgenus *Eremoctenia*. Legs long, tibiae long and slender, with several rows of setae in the distal part of the ventral margin.
Abdomen: Abdominal ctenidium either normally developed, reduced or absent. Postspiracular sclerite absent. Female abdomen with two or three tergites before the anal segment, sternites 5 and 6 with lateral sclerites or produced into processes. Dorsal and ventral genital plates present. Dorsal plate generally forming a lip above the genital opening, with long setae in some species. Tergites 1 and 2 of abdomen of male fused in some species.
Male Genitalia: Aedeagus usually curved, with ventral scale-like teeth and a ventral tooth at the apex in some species.
About twenty species distributed in temperate latudes and in the tropics of the Old World. Two species in Israel.

Penicillidia conspicua Speiser, 1901
Figures 324–329

Penicillidia conspicua Speiser, 1901. *Arch. Naturgesch.*, 67 : 36.
Nycteribia westwoodi Kolenati, 1856. *Die Parasiten der Chiropteren*, Brünn, p. 34.
nec *Nycteribia westwoodi* Guérin-Ménéville, 1844. *Iconographie du Règne Animal de Cuvier,* III, p. 556.
Megistopoda westwoodi Kolenati, 1857. *Wiener Ent. Mschr.*, 1 : 65.
Penicillidia westwoodi Kolenati. 1863. *Horae Soc. Ent. Ross.*, 2 : 69
Penicillidia conspicua Speiser. Theodor & Moscona (1954) *Parasitology*, 44 : 215.
Penicillidia conspicua Speiser. Theodor (1954) in Lindner, 66a, Nyteribiidae, p. 34; (1967) *Cat. Rothsch. Coll. Nyct.*, p. 358.

Length 3.5–4.0 mm. Colour brown.
Head: Four or five rows of setae between the eyes. Labella of proboscis shorter than theca.

* *Penicillidia conspicua* Speiser was designated as type of the nominate subgenus by Speiser, 1908, and has been wrongly given as type species of the genus by Theodor, 1967.

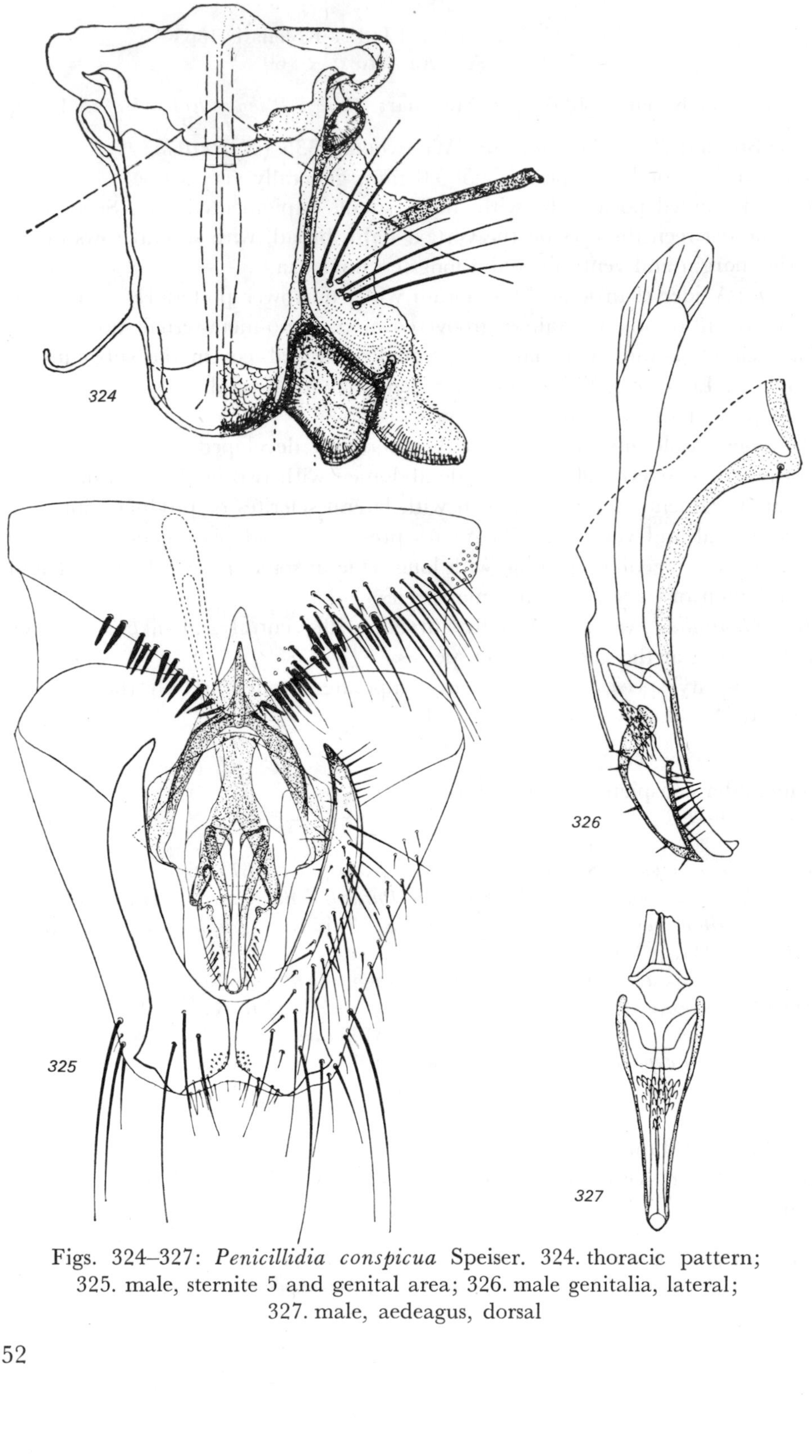

Figs. 324–327: *Penicillidia conspicua* Speiser. 324. thoracic pattern;
325. male, sternite 5 and genital area; 326. male genitalia, lateral;
327. male, aedeagus, dorsal

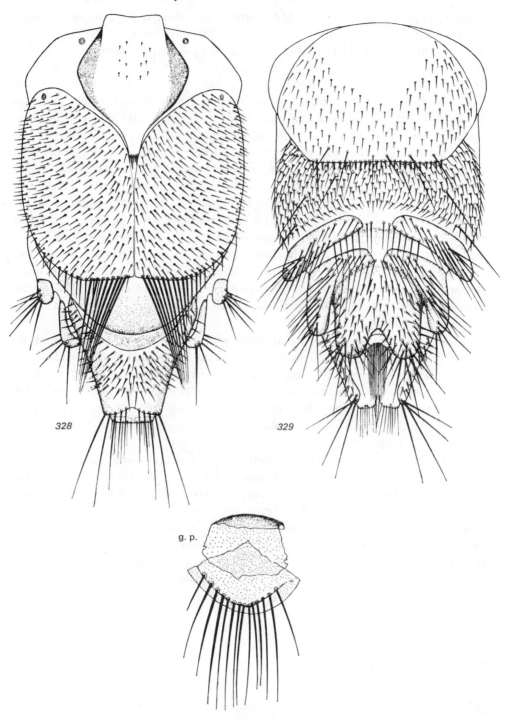

g. p.

328

329

Figs. 328–329: *Penicillidia conspicua* Speiser. Female abdomen.
328. dorsal, with genital plates (g.p.); 329. ventral

153

Thorax: Length to width = 3 : 4. Mesonotum wide, three to six notopleural setae, Lateral plates of notopleural sutures wide. Meso-metasternal sutures closed, forming an angle of 110°. Tibiae slender, scalpel-shaped, with three or four rows of setae at the distal part of the ventral margin.

Male Abdomen: Tergite 1 indistinctly divided from tergite 2. Tergite 2 with rounded posterior margin and a dense marginal row of moderately long and short setae. Tergites 3–6 with marginal rows of very long and shorter setae, particularly two to four very long setae in the middle of the rows. The tergites become shorter and more curved posteriorly. Surface of tergites 2–5 covered with short hairs, tergite 6 bare in its greater part. Sternite 1+2 rectangular, with a ctenidium of about forty ordinary spines, which are shorter in the middle and separated by narrow spaces. Sternite 5 triangular, with the apex posteriorly, its surface bare in the anterior part, with three or four rows of setae and about fifteen long spines in an irregular double row along the posterior margins, leaving the apex bare. *Genitalia*: Hypandrium triangular. Cerci curved, pointed, with numerous setae. Aedeagus curved, with a blunt ventral tooth at the apex, tapering, with scale-like teeth in the basal part of the ventral surface. Praegonites triangular, with curved ventral margin, pointed apical end and about ten setae at the dorsal margin.

Female Abdomen: Tergite 1 triangular, with two to six short spines at the apex, wedged into the large tergal plate 2. This is rounded posteriorly, incompletely divided in the middle and bears a row of fifteen to twenty long, closely placed setae in the middle of the posterior margin and shorter, more widely spaced setae laterally. The whole surface covered with short hairs. Tergal plate 3 absent. Anal segment conical, truncate, with concave anterior dorsal margin. Sternite 1+2 longer than in the male, with rounded sides and a ctenidium of thirty to thirty-five more widely spaced, ordinary short spines. Lateral sclerites of sternites 5 and 6 produced into posteriorly directed lobes with numerous setae. Sternite 7 bilobed posteriorly, also with numerous setae. Dorsal genital plate wide, triangular, with rounded posterior margin and a row of ten to fourteen long setae near the margin. Ventral plate wide, with rounded anterior margin.

Hosts: Species of *Miniopterus, Myotis, Rhinolophus.*

Distribution: S. Europe, S.W. Asia to Afghanistan; N. Africa.

Israel: Galilee; mainly on *Miniopterus schreibersi* (fifty specimens), *Myotis myotis* (seven specimens) and *Myotis mystacinus* (ten specimens); April to September, mainly in August and September.

Penicillidia dufourii (Westwood, 1835)
Figures 330–333

Nycteribia dufourii Westwood, 1835. *Trans. Zool. Soc. Lond.,* 1 : 275.
Nycteribia westwoodi Guérin-Méneville, 1844. *Iconographie du Règne Animal de Cuvier,* III, p. 556.
Nycteribia frauenfeldii Kolenati, 1856. *Die Parasiten der Chiropteren,* Brünn, p. 35.

Megistopoda leachii Kolenati, 1857. *Wiener Ent. Mschr.,* 1 : 62.

Penicillidia dufourii Westwood. Kolenati (1863) *Horae Soc. Ent. Ross.,* 2 : 72.

Penicillidia leachii Kolenati. Kolenati (1863) *ibid.,* p. 75.

Penicillidia dufourii Westwood. Speiser (1901) *Arch. Naturgesch.,* 67 : 32.

Penicillidia dufourii Westwood. Theodor & Moscona (1954) *Parasitology,* 44 : 211.

Penicillidia dufourii Westwood. Theodor (1954) in Lindner, 66a, Nycteribiidae, p. 36; (1967) *Cat. Rothsch. Coll. Nyct., p.* 362.

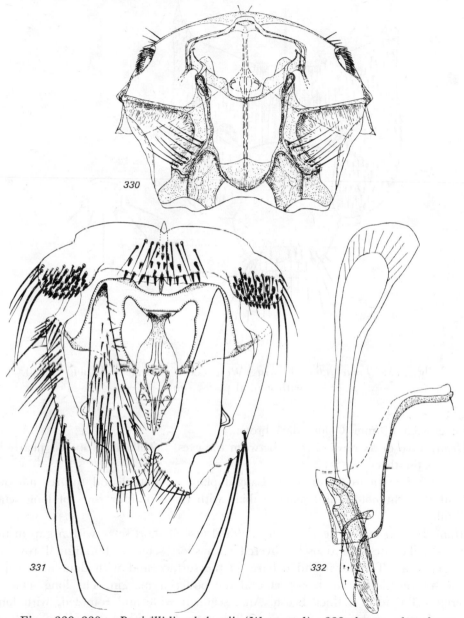

Figs. 330–332: *Penicillidia dufourii* (Westwood). 330. thorax, dorsal; 331. male, sternite 5 and genital area; 332. male genitalia

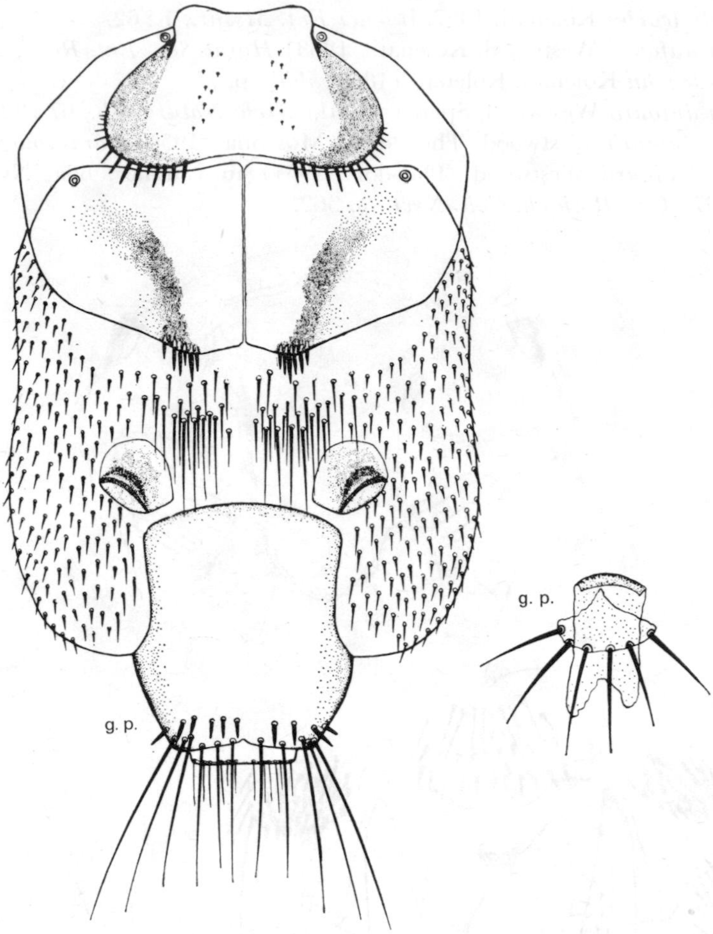

Fig. 333: *Penicillidia dufourii* (Westwood). Female abdomen, dorsal, with genital plates (g.p.)

Length 3.5-4.0 mm. Colour dark brown.

Head: Five to six rows of setae between the eyes. Labella of proboscis nearly as long as the theca.

Thorax: Length to width 3 : 4. Lateral plates of notopleural sutures narrow. Four to eight notopleural setae. Tibiae with three or four rows of long setae distally.

Male Abdomen: Tergite 1 with a marginal row of short setae with a gap in the middle. Tergites 2–4 densely covered with short setae and marginal rows of longer setae. Tergites 5 and 6 bare on the surface and with double or triple rows of very long and strong setae at the posterior margin. The long setae of tergites 4–6 form a thick brush. Anal segment wide and rounded, with long curved setae posteriorly. Sternite 1+2 rectangular, with a ctenidium of forty to forty-four spines which stand more closely than in *P. conspicua*. Sternite 5 with broad, lateral processes with flat surface, which are densely covered with short, thick spines. A triangular group of similar spines in the middle of the sternite. There is a gap between the median group and the spines on the lateral processes. *Genitalia*: Hypandrium nearly square, with broad lateral extensions. Cerci thick, straight, tapering to a blunt, dark point, densely covered with long setae in the basal part and shorter setae apically. Aedeagus straight, with rounded end and without scales on the ventral surface. Praegonites narrow, triangular, with sharp apical point and three or four setae at the dorsal margin.

Female Abdomen: Tergite 1 wide, with a marginal row of short setae with a large gap in the middle. Tergal plate 2 about three times as wide as long, bare on the surface and incompletely divided in the middle. Two groups of three to nine long spines near the middle of the posterior margin. Pigmented stripes run from these spines obliquely towards the anterior margin. Several rows of setae on the connexivum behind tergal plate 2. Two sclerotized processes with sharp, pigmented posterior margin which project posteriorly lateral and posterior to these setae. These processes are probably remnants of lateral sclerites of tergal plate 3. Anal segment large, rounded, bare on the surface, except for a row of spines and long setae posteriorly. Pleurae covered with short hairs. Sternite $1+2$ rectangular, with a ctenidium of forty to forty-four closely-standing spines. Sternites 5 and 6 with lateral sclerites which are directed posteriorly, forming lobes. Sternite 7 with two similar lobes. All lobes of sternites 5–7 with brushes of setae. Dorsal genital plate small, triangular, with five to seven setae at the posterior margin. Ventral plate narrow, with rounded anterior magin.

Hosts: Species of *Myotis*, *Miniopteus*, *Rhinolophus* and other general.

Distribution: Continental Europe; S.W. Asia to the Himalayas; N. Africa.

Israel: Galilee; eighty-five specimens on *Miniopterus schreibersi*, sixty-three specimens on *Myotis myotis*, thirty-one specimens on *M. nattereri* and thirty specimens on *M. mystacinus;* May to October, mainly in August.

Subfamily CYCLOPODIINAE

DIAGNOSIS

Parasites of Megachiroptera. Head laterally or dorso-ventrally compressed. Eyes always present. Mesonotum widening posteriorly. Mesopleural sutures originating far posteriorly, except in *Archinycteribia*. Tibiae cyclindrical, with two or three bands of weaker integument and short setae in the middle. Notopleural setae usually reduced in number, rarely absent. Tergites 1 and 2 or abdomen fused in all species in both sexes. Segmentation of abdomen of female markedly reduced. Setae coarse, spines of ctenidia coarse and blunt. Genitalia of *Nycteribia*, *Eucampsipoda* and *Cyclopodia* type. Old World only. Five genera, of which only one is Palaearctic and occurs in Israel.

Diptera Pupipara

Genus EUCAMPSIPODA Kolenati, 1857
Wiener Ent. Mschr., 1 : 62.

Type Species: *Nycteribia hyrtlii* Kolenati, 1856.

Head: Laterally compressed, pigmented dorsally. Eyes single-faceted. Labella of proboscis markedly longer than theca.

Thorax: Pentagonal. Thoracic and abdominal ctenidia consisting of thick, blunt spines. Mesonotum narrow, widening posteriorly. One long and one very short notopleural seta. Haltere groove without cover. Meso-metasternal sutures open. Tibiae long, parallel-sided, nearly cylindrical, with two rings of weaker integument and short setae in the middle.

Abdomen: Postspiracular sclerite with a single seta. Segmentation of abdomen of female markedly reduced, only tergite 6 and sternite 6 present between the basal sclerites and the anal segment. Ventral genital plate (sternite 7) of female divided into two sclerites with spines.

Genitalia: Aedeagus of male tubular, with a large endophallus, connected with the phallobase by a long connective membrane which bears numerous spines. Cerci straight, thin, tapering, covered with peg-like spines and setae.

Twelve species, mainly distributed in the tropics of the Old World, one species in the Middle East.

Eucampsipoda hyrtlii (Kolenati, 1856)
Figures 232, 239, 250, 253–255, 263, 267, 334–345

Nycteribia hyrtlii Kolenati, 1856. *Die Parasiten der Chiropteren,* Brünn, p. 42.
Nicteribia fitzingeri Kolenati, 1856. *Ibid.,* p. 43.
? *Nycteribia aegyptia* Macquart, 1851. *Mém. Soc. Sci. Agric. Lille,* p. 282.
Eucampsipoda aegyptiaca Macquart. Kolenati (1863) *Horae Soc. Ent. Ross.,* 2 : 80.
Eucampsipoda hyrtlii Kolenati. Theodor (1954) in Lindner, 66a, Nycteribiidae, p. 40; Theodor & Moscona (1954) *Parasitology,* 44 : 218; Theodor (1955) *ibid.,* 45 : 195; (1967) *Cat. Rothsch. Coll. Nyct.,* p. 413.

Length 2.5–3.0 mm. Colour reddish brown.

Head: Dark dorsally. Eyes large, situated close to the anterior margin. Labella of proboscis twice as long as the theca. Palps thick at the base, slender apically, with a long terminal seta.

Thorax: As for the genus. Notopleural sutures close together anteriorly, markedly diverging posteriorly. Meso-metasternal sutures slightly curved, forming a right angle. Thoracic ctenidia with sixteen to eighteen curved, blunt spines. One long and one very short notopleural seta.

Male Abdomen: Tergites 1 and 2 fused, trapezoidal, with rounded posterior corners. Tergites 3-6 short. Tergites 2-6 with moderately long setae at the posterior margin which are longer on the posterior tergites, particularly two to four setae in the middle of the marginal rows of tergites 5 and 6. Abdominal

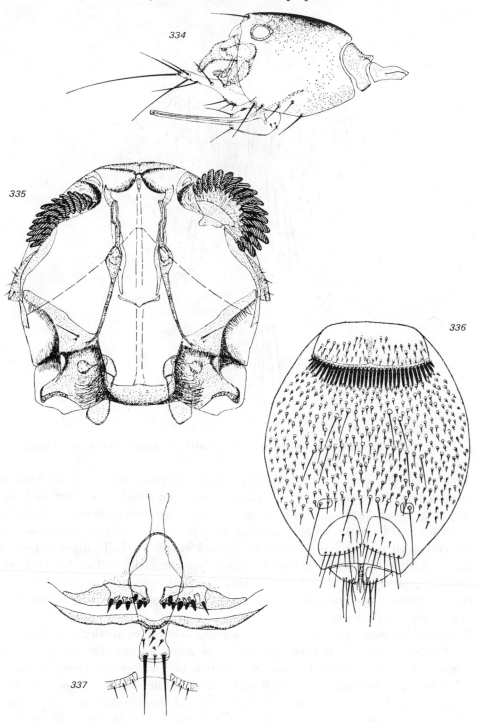

Figs. 334–337: *Eucampsipoda hyrtlii* (Kolenati). 334. head, side view; 335. thorax, dorsal; 336. female abdomen, ventral; 337. female, genital plates

159

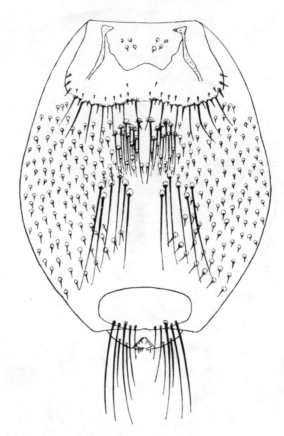

Fig. 338: *Eucampsipoda hyrtlii* (Kolenati). Female abdomen, dorsal

ctenidium with about thirty thick, blunt, flattened spines. Sternite 5 triangularly produced in the middle of the posterior margin and with two short and two very short spines at the apex, which may be absent in some specimens. *Genitalia*: Cerci long, 0.43 mm, straight, tapering to a point in dorsal view, widened at the end, with a short ventral tooth at the end in side view. Two rows of short, thick, peg-like spines on the dorsal surface from near the end and a long seta dorsally near the base. Hypandrium triangular. Aedeagus nearly cylindrical, with oblique anterior opening, 0.43 mm long, eight times as long as wide at the base. Praegonites small, triangular.

Female Abdomen: Tergite 1+2 with a marginal row of moderately long and short setae. Connexivum with two groups of setae, an anterior group close to tergite 2 with about ten thin setae in each half and a posterior group of five to seven long and strong setae in each half in the middle of the dorsum. The area between and behind the long setae bare. Sides of dorsum and pleurae covered with short spines. Tergite 6 short, transvrse-elliptical, with eight to ten long setae at the posterior margin. Anal segment very short. Sternite 1+2 shorter than in the male. Abdominal ctenidium with thirty-two to thirty-six thick spines.

Sternite 6 divided into triangular lateral sclerites with a marginal row of long and short setae. Ventral genital plate (sternite 7) divided into wing-shaped halves with about six short pegs in each half. Dorsal genital plate longer than wide, rounded anteriorly, fused with the anal sclerite which is broad and bears two setae posteriorly and a few short spines on the surface.

Hosts: *Rousettus aegyptiacus* Geoffroy, *Eidolon sabaeum* Andersen.

Distribution: Middle East, Syria to Arabia; possibly E. Africa.

Israel: Galilee to southern Coastal Plain, Jerusalem; common on *Rousettus aegyptiacus* throughout the year.

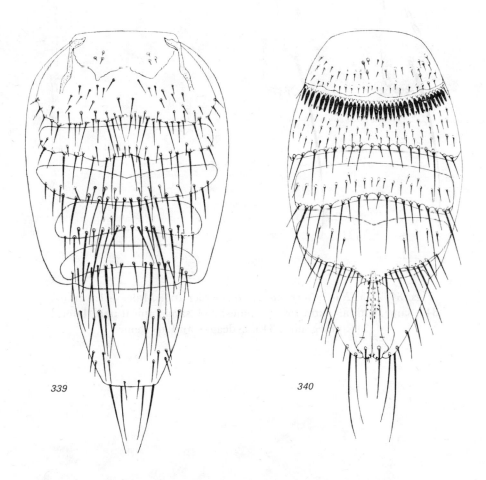

Figs. 339–340: *Eucampsipoda hyrtlii* (Kolenati). Male abdomen. 339. dorsal; 340. ventral

161

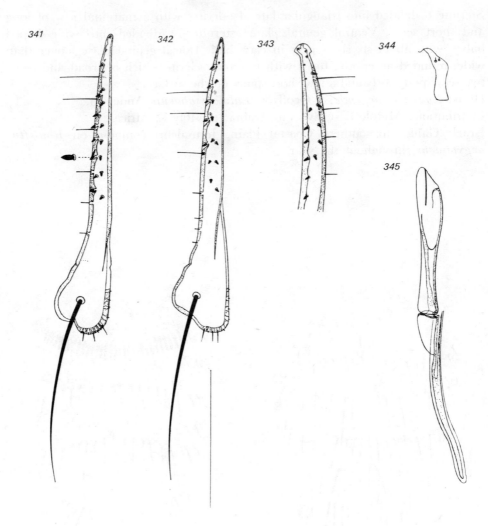

Figs. 341-345: *Eucampsipoda hyrtlii* (Kolenati). Male. 341. cercus; 342. same, with an extra row of spines; 343. same, apical part, lateral; 344. praegonite; 345. aedeagus and apodeme

REFERENCES

Bequaert J. (1942) 'A Monograph of the Melophaginae or Ked-Flies of Sheep, Goats, Deer and Antelopes', *Entomologica Americana,* NS 22 : 1–220.

— (1953/1957) 'The Hippiboscidae or Louse Flies of Mammals and Birds', *ibid.,* NS 1 : 1–442; 2 : 1–611.

Hoare C. (1923) 'An Experimental Study of the Sheep Trypanosome (*T. melophagium* Flu) and its Transmission by the Sheep Ked *(Melophagus ovinus* L.)', *Parasitology,* 15:365–424.

Hurka K. (1964) 'Distribution, Bionomy and Ecology of the European Bat Flies with Special Reference to the Czechoslovak Fauna (Nycteribiidae)', *Acta Universitatis Carolinae* (Biologica), pp. 157–234.

Hurka K. & D. Povolny (1968) 'Faunal and Ecological Studies on Nycteribiidae and Streblidae of Nangrahar Province (Eastern Afghanistan)', *Acta Entomologica Bohemo–Slovaca,* 65 : 285–298.

Jobling B. (1926) 'A Comparative Study of the Head and Mouth Parts in the Hippoboscidae', *Parasitology,* 18 : 319–349.

— (1928) 'The Structure of the Head and Mouth Parts in the Nycteribiidae', *ibid.,* 20 : 254–272.

— (1929) 'A Comparative Study of the Head and Mouth Parts in the Streblidae', *ibid.,* 21 : 417–445.

— (1939) 'On the African Streblidae Including the Morphology of the Genus *Ascodipteron* and Description of a New Species', *ibid.,* 31 : 147–165.

Kemper H. (1951) 'Beobachtungen an *Crataerina pallida* und *Melophagus ovinus*', *Zeitschrift für hygienische Zoologie und Schädlingsbekämpfung,* 39 : 225–259.

Kolenati F. A. (1863) 'Beiträge zur Kenntnis der Phthiriomyiarien', *Horae Societatis Entomologicae Rossicae,* 2 : 9–118.

Maa T. C. (1963) *Genera and Species of Hippoboscidae — Types, Synonymy, Habitats and Natural Groupings (Pacific Insects Monograph,* VI), pp. 1–186.

— (1965a) 'A Synopsis of the Lipopteninae', *Journal of Medical Entomology,* 2 : 233–248.

— (1965b) 'Ascodipterinae of Africa', *ibid.,* 2 : 311–326.

— (1966) *Studies in Hippoboscidae,* Part 1 *(Pacific Insects Monograph,* X), pp. 1–148.

— (1969) *Studies in Hippoboscidae, Part 2 (Pacific Insects Monograph,* XX), pp. 1–312.

— (1971 *Studies in Bat Flies (Streblidae, Nycteribiidae), (Pacific Insects Monograph,* XXVIII), pp. 1–248.

Schlein Y. (1967) 'Postmetamorphic Growth in *Lipoptena capreoli* Rondani and other Insects', *Israel Journal of Zoology,* 16 : 69–82.

— (1968) 'The Thoracic Skeleton and Musculature of Winged and Wingless Pupiparous Flies (Diptera)', Ph.D. Thesis, The Hebrew University of Jerusalem (in Hebrew).

— (1970) 'A Comparative Study of the Thoracic Skeleton and Musculature of the Pupipara and the Glossinidae', *Parasitology,* 60 : 327–373.

Schlein Y. & O. Theodor (1971) 'On the Genitalia of the Pupipara and their Homologies with those of *Calliphora* and *Glossina*', *ibid.,* 63 : 331–342.

Speiser P. (1900) 'Über die Strebliden', *Archiv für Naturgeschichte,* 66 : 31–70.

— (1901) 'Über die Nycteribien', *ibid.,* 67 : 11–78.

— (1908) 'Die geographische Verbreitung der Diptera Pupipara und ihre Phylogenie', *Zeitschrift für Wissenschaftliche Insektenbiologie,* 4 : 241–246; 301–305; 420–427; 437–447.

Theodor O. (1928) 'Über ein nicht pathogenes *Trypanosoma* aus der Ziege und seine Übertragung durch *Lipoptena caprina* Austen', *Zeitschrift für Parasitenkunde,* 1 : 283–300.

— (1954) 'Nycteribiidae', in: E. Lindner (ed.) *Die Fliegen der Palaearktischen Region,*
— 66a : 1–44.

— (1954) 'Streblidae', *ibid.,* 66b : 1–12.

— (1955) 'On the Genus *Eucampsipoda* and *Dipseliopoda* n. gen.', *Parasitology,* 45 : 195–229.

References

— (1957) 'Parasitic Adaptation and Host-Parasite Specificity in the Pupiparous Diptera', I. *Symposium on Host Specificity among Parasites of Vertebrates*, Neuchâtel. pp. 50–63.

— (1963) 'Über den Bau der Genitalien bei den Hippobosciden', *Stuttgarter Beiträge zur Naturkunde*, 108 : 1–15.

— (1967) *An Illustrated Catalogue of the Rothschild Collection of Nycteribiidae* (Diptera) *in the British Museum (Natural History)*, London, pp. 1–506.

— (1968) 'A Revision of the Streblidae of the Ethiopian Region', *Transactions Royal Entomological Society London*, 120 : 313–373.

Theodor O. & A. Moscona (1954) 'On Bat Parasites in Palestine, I : Nycteribiidae, Streblidae, Hemiptera, Siphonaptera', *Parasitology*, 44 : 157–245.

Theodor O. & H. Oldroyd (1964) 'Hippoboscidae', in: E. Lindner (ed.), *Die Fliegen der Palearktischen Region*, 65 : 1–70.

164

INDEX

Synonyms in italics. The principal reference to each valid name in bold type.

Index

166

Index

Geographical Areas in Israel and Sinai

KEY

1. Upper Galilee
2. Lower Galilee
3. Carmel Ridge
4. Northern Coastal Plain
5. Valley of Yizre'el
6. Samaria
7. Jordan Valley and Southern Golan
8. Central Coastal Plain
9. Southern Coastal Plain
10. Foothills of Judea
11. Judean Hills
12. Judean Desert
13. Dead Sea Area
14. ʿArava Valley
15. Northern Negev
16. Southern Negev
17. Central Negev
18. Golan Heights
19. Mount Hermon
20. Northern Sinai
21. Central Sinai Foothills
22. Sinai Mountains
23. Southwestern Sinai

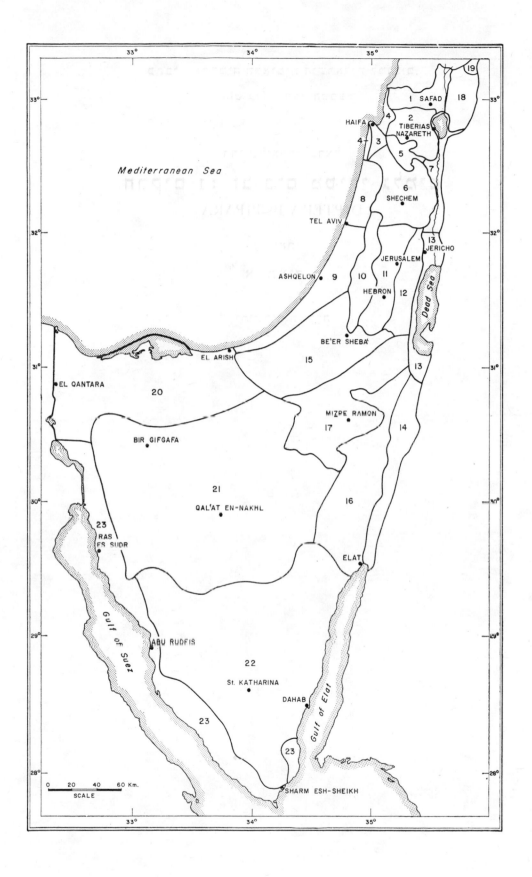

Mediterranean Sea

33°

1 SAFAD
18
19

HAIFA
4
2
TIBERIAS
NAZARETH
4
3
5
7
6
SHECHEM
8
TEL AVIV
13
JERICHO
JERUSALEM
ASHQELON
9
10
11
HEBRON
12
Dead Sea
BE'ER SHEBA'
13
EL ARISH
15
EL QANTARA
20
MIZPE RAMON
17
14
BIR GIFGAFA
21
QAL'AT EN-NAKHL
16
23
RAS ES SUDR
ELAT
Gulf of Suez
ABU RUDFIS
22
St. KATHARINA
DAHAB
Gulf of Elat
23
23
SHARM ESH-SHEIKH

0 20 40 60 Km.
SCALE

כתבי האקדמיה הלאומית הישראלית למדעים

החטיבה למדעי-הטבע

החי של ארץ-ישראל

חרקים 1: זבובים מטילי גלמים

(DIPTERA PUPIPARA)

מאת

א' תאודור

ירושלים תשל"ה